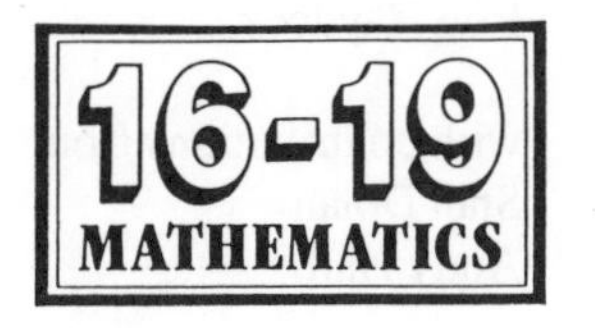

Information and coding

The School Mathematics Project

Main author Ron Haydock

with contributions from
Stan Dolan
Andy Hall

Project director Stan Dolan

The authors would like to give special thanks to Ann White for her help in preparing this book for publication.

The publishers would like to thank the following for supplying photographs:

Page 4 - Ron Haydock
Page 66 - NASA/Science Photo Library.

Cartoons by Paul Holland and Margaret Ackroyd
Illustration on page 53 by Paul Brown
Illustrations on pages 60, 61 and 62 by Hardlines

Published by the Press Syndicate of the University of Cambridge
The Pitt Building, Trumpington Street, Cambridge CB2 1RP
40 West 20th Street, New York, NY 10011-4211, USA
10 Stamford Road, Oakleigh, Victoria 3166, Australia

First published 1991
Reprinted 1992

Produced by 16-19 Mathematics, Southampton

Printed in Great Britain by Scotprint Ltd., Musselburgh.

ISBN 0 521 42647 2

Contents

1. What is information?

1.1 Introduction

In this unit you will find only a brief introduction to an important and rapidly growing subject. The theory of information and coding is used chiefly in computing science but has relevance for many other areas of interest, ranging from linguistics to animal behaviour. Suggestions for further reading are given in the *Unit Guide*, together with some topics and themes for investigation.

TASKSHEET 1 - *Non-verbal information*

1.2 Communication

In a telephone conversation, two people communicate ideas or thoughts to each other. The thoughts of the speaker are expressed in spoken words and the sounds of the words are converted by the mouthpiece into an electric signal. This signal is conveyed along a wire and reconstituted into sound by the earpiece of the listener, who interprets the sound as words, with meaning.

This typical example of communication involves a two-stage encoding:

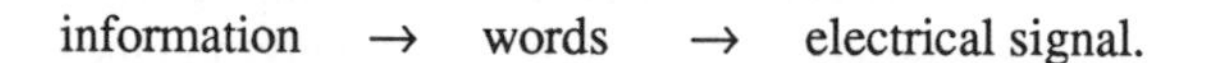

information → words → electrical signal.

The encoding is reversed at the receiving end in the decoding procedure. Any communication process may be represented diagrammatically:

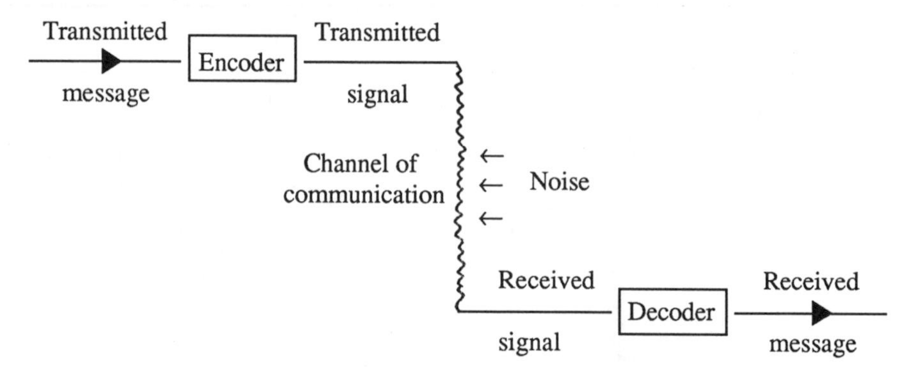

Note that, in this instance, codes are not used to keep a message secret, as in a spy story. In this unit, codes will be used to make information **more** rather than less accessible. You might define an encoding as an alternative representation of a message, making it more suitable for the purpose of signalling.

The word '**noise**' in the diagram denotes anything which causes a difference between the transmitted and the received message, ranging from random (Brownian) motion of the electrons in the telephone wire to the mishearing of a word.

(a) Identify as far as you can the message, the encoding procedure and the 'noise' in

(i) a conversation at a party,

(ii) a tic-tac communication at a racecourse,

(iii) exchange of information amongst bees about the whereabouts of suitable flowers.

(b) How might you measure the amount of information in each case?

In this book you will discover how to measure information. You will go on to consider the problem of noise and how to minimise its adverse effects.

1.3 Binary encoding of information

Suppose that you and your friend play a guessing game. She has four solid objects - a cube, a sphere, a cone and a square-based pyramid. She chooses one of them, and you discover which by asking questions to which she will answer only 'yes' or 'no'.

What questions might you ask? How many questions will you need to ask?

In the most refined method, you should find that two questions are always enough. The trick is to halve the remaining possibilities at each guess. Thus the first question might be: 'Has it a curved face?'

The sequence of questions used to identify the object might then be represented by the **binary decision tree:**

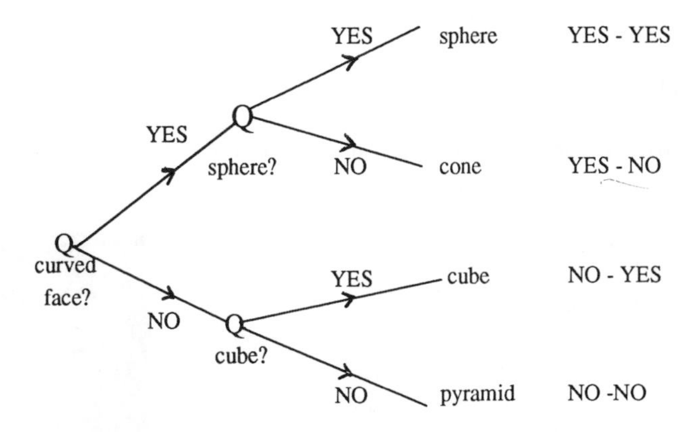

The tree is called **binary** because two branches are directed outward from each node except the tips. If you use this tree in all such games, then the four objects are in effect **labelled,** as shown above.

Since only two symbols are used, these are binary labels. Replacing 'yes' by 1 and 'no' by 0 you have:

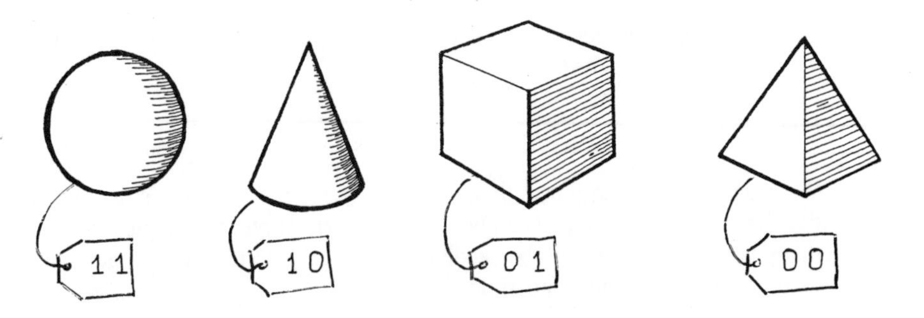

Exercise 1

1. A friend has a set of eight cards as shown.

   (a) Show that three yes/no type questions will identify a particular card with certainty.

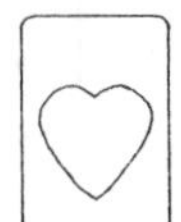

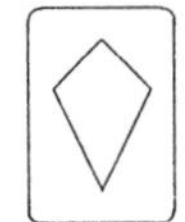

   (b) In fact, the same three questions, in the same order each time, will succeed in every game. What might the questions be?

   (c) Draw a binary decision tree based on your set of questions and allocate binary labels to the eight cards.

2. An ornithologist observes a flock of birds, noting for each individual whether it is
   A : male or female,
   B : mature or immature,
   C : in full plumage or not,
   D : on the ground or perching.

   (a) How many classes of bird will be identified in her notes?

   (b) Suggest a binary labelling and use it to label a mature hen in full plumage perching in a tree.

3. A sign painted on a stone in Derbyshire is shown in the photograph.

   (a) (i) Assuming that this is part of a numbering system, how would you extend it to represent the numbers 0-15?

   (ii) In what sense is your system a binary system?

   (b) If the sign is not part of a number system, what other meaning might it have?

4. How many instructions to a computer are possible if each instruction consists of eight binary digits?

Possible project

Write a convincing argument that in the first game the 'halving' strategy used to identify the solid is the best that can be devised. (Which other strategies are there and how many yes/no questions will be needed on average?) Extend your argument to the eight card game and beyond, if possible.

1.4 Measuring information

It is conventional to measure **information** in binary digits (abbreviated to **bits**). For example, the information capacity of computer memories is measured in K or thousands of bits. A 128K memory is one which can store 128 000 binary digits.

Before your guesses in the four solids game you knew that the answer was one of four equally likely objects. You were able to label the four objects using two binary digits as 00, 01, 10 and 11. Finding which object your friend chose is therefore the same as determining two binary digits or acquiring two bits of information.

In *Information Theory* the set of all possible objects which might be chosen is called the **alphabet** or **source**, often denoted by S. The information conveyed by determining a specific member of the source, the **information per member**, is denoted by $H(S)$. So when

$$S = \{\text{sphere, cone, cube, pyramid}\},$$

$$H(S) = 2.$$

You have seen that:

$$n(S) = 4 \Rightarrow H(S) = 2 \text{ (the solids game)},$$

$$n(S) = 8 \Rightarrow H(S) = 3 \text{ (the cards game)}$$

If $n(S) = 2^k$, what is $H(S)$? Justify your answer.

TASKSHEET 2S - *Indices and logarithms*

If a set S has 2^k equally likely members, then k bits are needed to specify each member. This is used as a basis for a general definition:

If a source S has n equally likely members then the 'theoretical' information per member is defined to be the power of 2 equal to n, i.e.

$$H(S) = \log_2 n.$$

A table of $\log_2$ (Table 1) is provided at the end of the unit.

When $n(S)$ is a power of 2, you have seen that only H(S) yes/no questions are required to determine a particular member of S. When $n(S)$ is not a power of 2 you may not be able to attain this theoretical minimum.

Consider yet another guessing game, that of finding which of the numbers 1-10 is being 'thought of'. The source is therefore {1, 2, 3, 4, 5, 6, 7, 8, 9, 10}. Assume that numbers are all chosen with the same frequency, i.e. that they are equally likely. You can start a decision tree with a question that halves the source:

'Is it greater that 5?'

After that, the nearest you can get to the 'halving' strategy is shown in this decision tree:

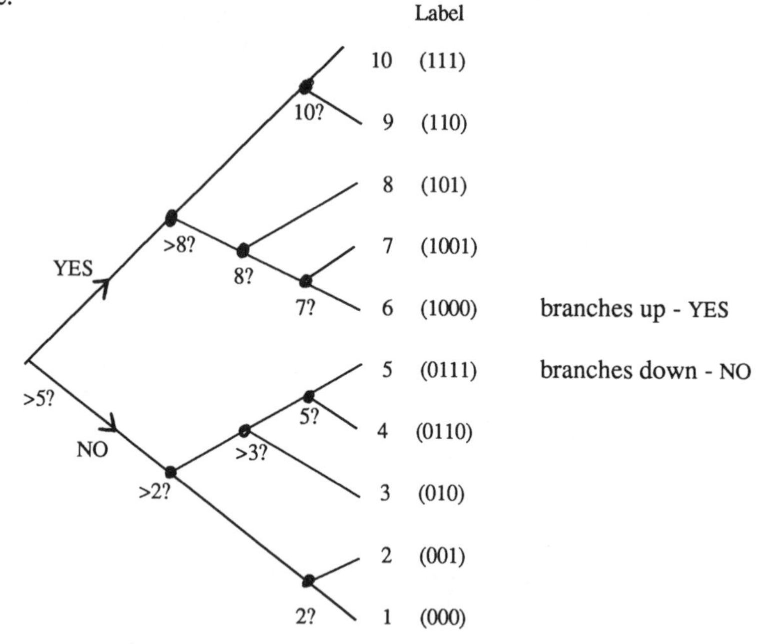

In six cases you need three questions and in the remaining four cases, four questions. So the average number of questions is

$$\frac{(6 \times 3) + (4 \times 4)}{10} = 3.4 ,$$

whereas H(S) is $\log_2 10 = 3.322$.

(a) Find H(S) if $n(S) = 12$.

(b) What is the average number of questions needed to find somebody's month of birth?

(c) Suggest a possible set of questions.

Example 1

A combination bicycle lock consists of four drums, each of which may be rotated independently on the same axis. Each drum has the figures 0 to 9 on it. The lock opens when the correct four figures are in line in the correct position. How many bits of information do you need to open the lock?

Solution

There are 10000 possible combinations, consisting of all the numbers from 0000 to 9999. Assuming these are equally likely, the information content is $\log_2 10000 = 13.3$ bits.

The solution above was simple and straightforward. However, there is another way of looking at the problem which should give the same answer:

Each of the four digits has uncertainty $\log_2 10$. The total information content is therefore $\log_2 10 + \log_2 10 + \log_2 10 + \log_2 10 = 4 \log_2 10$. This is the same as $\log_2 10000$ because, using a property of logarithms,

$$\log_2 10000 = \log_2 10^4 = 4 \log_2 10.$$

The fact that the answers are the same illustrates the appropriateness of the '$\log_2 n$' definition of information content. The idea can be generalised:

If the symbol s_1 carries information c_1 bits, symbol s_2 has information content c_2 bits and so on, then the string $s_1 s_2 \ldots s_n$ has information content

$$(c_1 + c_2 \ldots + c_n) \text{ bits.}$$

For the result above, it is assumed that the members of the string $s_1 s_2 \ldots s_n$ are independently chosen, that is the choice of one member does not affect the likelihood of the choice of another.

Exercise 2

1. One of the 26 letters of the alphabet is chosen at random and you have to guess which by using yes/no questions.

   (a) What is the least number of questions necessary to be **certain** of finding the chosen letter?

   (b) Devise a strategy to minimise the average number of questions. What would be that average number?

   (c) What is the information content of a member of the source?

2. Without drawing a tree, obtain the average number of questions needed to find one of the numbers 1 to 100.

3. Suppose that 27 Scrabble tiles representing the 26 letters and a space were put into a hat, drawn at random one at a time, noted and replaced.

   (a) What would be the average length of a 'word' (i.e. the sequence of letters between two spaces)?

   (b) How many yes/no questions would be needed on average to identify an element of the source?

   (c) How many such questions would you need on average to guess a five-letter word?

   (d) What is the (theoretical) information content of a five-letter word?

4. A black and white TV monitor displays approximately 600 rows and 800 columns of dots, each of which may have any one of 10 degrees of brightness. Calculate the information content of a picture on the screen. An old saying has it that a picture is worth a thousand words. What do you think?

Possible project

A display board at a football stadium consists of a grid of rectangular glass panels.

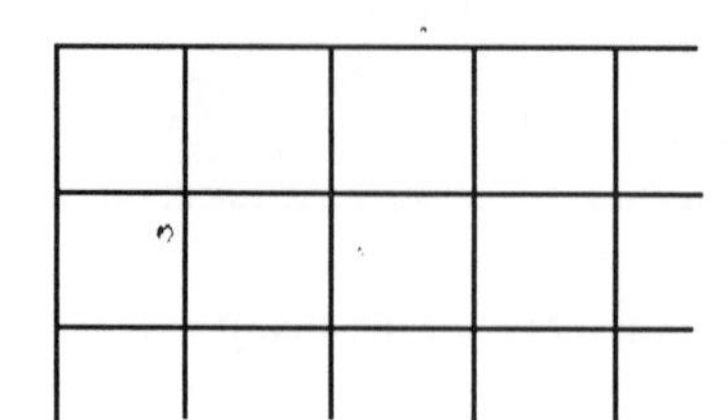

Behind each panel is the same arrangement of neon tubes. Using on-off switches, a controller can light up a selection of the tubes in each panel in turn, so that the whole display spells out a message.

(a) Design a suitable arrangement of tubes for each panel.

(b) Explain and discuss the system.

After working through this chapter you should:

1. appreciate that all communication involves encoding and decoding procedures;

2. appreciate the wide range of applications of Information Theory;

3. understand and be able to use the terms **information** and **source**;

4. for a source with equally likely members, be able to obtain an approximation to the information content of a member by constructing a binary decision tree;

5. understand why, for such a source S with n members,

$$H(S) = \log_2 n$$

and be able to apply this result to find the total information content of a sequence of members of S.

Non-verbal information

A short walk in any town will reveal dozens of signs, familiar and unfamiliar, which convey information without the use of words. Many of them are tangible objects, such as road signs, the registration plates of cars and the signs used to indicate water hydrants. There are intangible indicators too, many of which are categorised as 'body language'; you can recognise at a glance the girl who is waiting for her boyfriend or the person who is timid or afraid or the individual who feels the need of a big 'personal space'. And your mere glance conveys to **them** your interest in their affairs.

Other environments offer yet more signs; there are the specialised indicators used on the railways or by the canals. In the country, you can see signposts, footpath and forestry signs, and other less tangible signs such as flattened crops indicating wind damage.

(Continued)

Some signs require close study or special equipment or specialised knowledge before they yield their information content. By studying packets and tins of food you might discover how the retail trade bar code works. By careful inspection of the phosphor spots on envelopes, used to represent postal codes, you might work out the connection between dots and corresponding letters and numbers. Either of these codes is read immediately by the appropriate scanning equipment, at the supermarket or post office. Similarly, a cash dispenser is able to extract the information on a bank or credit card.

The classification of signs and symbols is too big a subject to be attempted here, but some examples among street signs may be of interest. Many of them are **house marks**, like the logos of banks, or the barber's pole; they identify premises of particular trades or professions. Some are simply pictorial representations of words; inn-signs for the Bull or the White Lion are examples.

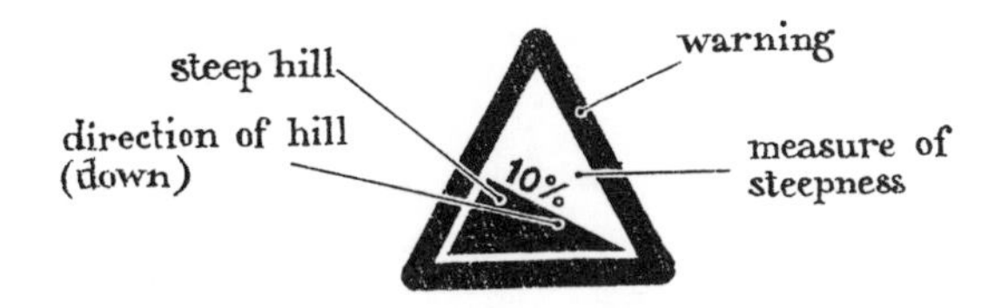

Road signs often incorporate a picture of a word or idea together with a conventional code (using the shape of the sign) and possibly some numerical information.

A particularly interesting device is the rebus or puzzle sign an example of which is the house mark of the publisher Hart Davis, which incorporates a picture of a hart. (If the picture were a **heart** it would be called a **phonetic** rebus.) An example of a phonetic rebus is the six cowrie shells offered by a Yoruba tribesman to declare his interest in a girl. The Yoruba word for six, *efa*, suggests the word for attraction, *fa*. In reply she might send eight shells, the word for eight, *eja*, signifying agreement, *jo*.

Possible project

Find out how the bar codes used in the retail trade work. A reference is given in the unit guide, together with other suggested themes for extended investigation. Having selected a theme you can start collecting material related to it.

Indices and logarithms

WHAT IS INFORMATION?

TASKSHEET 2S

Since logarithms are used a good deal in this book it will be as well to revise their definition and main properties. Logarithms are merely 'indices brought down to earth' so you should start with revision of the properties of numbers written with indices ('exponentially').

If a is a number greater than zero and m and n are *any* numbers,

(1) $a^0 = 1$

(2) $a^m \times a^n = a^{m+n}$

(3) $a^m \div a^n = a^{m-n}$

(4) $(a^m)^n = a^{mn}$

1. Express in a form without indices:

(a) 143^0 (b) $25^{\frac{1}{2}}$ (c) 4^{-2}

(d) 3^{-1} (e) $64^{\frac{2}{3}}$ (f) $16^{-\frac{3}{4}}$

(g) $4^{2.5}$ (h) $27^{-\frac{4}{3}}$

2. Solve the equations

(a) $3^x = 243$ (b) $5^x = 1$ (c) $2^x = \frac{1}{4}$

(d) $9^x = 27$ (e) $4^x = \frac{1}{8}$ (f) $(\frac{1}{8})^x = 4$

You probably recall common logarithms and remember that because, say, $10^3 = 1000$ then $\log_{10} 1000 = 3$. In general,

$n = b^r \Leftrightarrow \log_b n = r$

The positive number b is called the base of the logarithm r.

(Continued)

For *Information and coding*, logarithms to base 2 are especially important.

$$n = 2^r \Leftrightarrow \log_2 n = r.$$

From the results for indices it is possible to formulate the rules:

For any positive numbers b, m and n and any number r,

(1) $\log_b 1 = 0$

(2) $\log_b m + \log_b n = \log_b mn$

(3) $\log_b m - \log_b n = \log_b (\frac{m}{n})$

(4) $r \log_b m = \log_b m^r$

3. Use the rules to simplify

(a) $\log_2 (8 \times 64)$ (b) $\log_2 (1 \div 2)$ (c) $\log_2 (4^4)$

(d) $\log_2 (\frac{1}{8})$ (e) $\log_2 (\frac{1}{4})$ (f) $2 \log_2 3 + \log_2 4$

4. Solve the equations

(a) $\log_2 x = 4$ (b) $\log_2 x = \frac{1}{2}$

From the definition of a logarithm stated earlier you should note that:

if $f(x) = 2^x$ $(x > 0)$ then $f^{-1}(x) = \log_2 x$.

Logarithmic functions are the inverses of exponential functions.

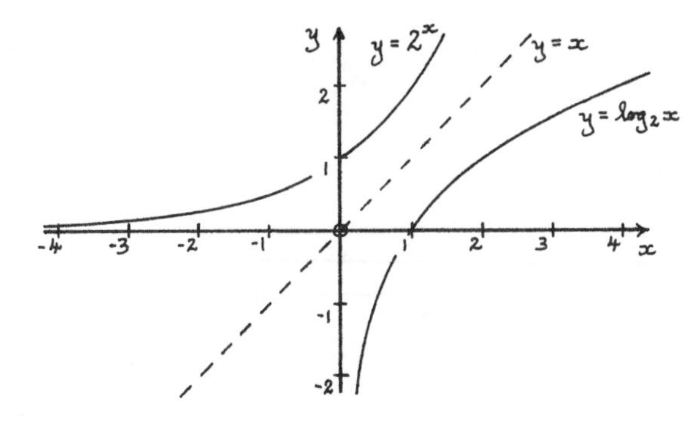

The graph of $y = \log_2 x$ is the reflection in $y = x$ of the graph of $y = 2^x$; the domain of a logarithmic function is the positive numbers.

Tutorial sheet

WHAT IS INFORMATION?

TASKSHEET 3

1. The following reference number was given to a client by a motoring organisation.

(a) Using 0 and 1 instead of space and bar, and reading from top to bottom in each column, the decimal digit 2 has been coded as 0011. Find the codenames for each of the given decimal digits.

(b) Explain why the binary codenames have four digits.

(c) What is the total information content of a decimal number with 18 digits?

(d) Disregarding the 'spaces', how many binary digits are used to represent the reference number?

2. A facsimile (FAX) transceiver prepares a signal as follows. An A4 page containing text and/or drawings is scanned line by line. The machine works in terms of white and black only, the scanner noting in each position the presence or absence of white paper. Samples are taken at 1728 equally spaced positions on each line. The quality of reproduction required determines the number of lines per mm, which is

3.85 for standard reproduction, 7.7 for fine, 15.4 for superfine.

(a) If the sample were equally likely to be black and white in each case, what is the information content of a superfine FAX sheet?

(b) Explain why black and white are **not** equally likely.

2 *Source encoding*

2.1 Unequal frequencies

So far you have considered the members of a source to be equally likely. This is not always so. For example, Tables 2 and 3 indicate the range of frequencies of letters and some words, respectively, in the English language. From Table 2 you can see that in a piece of English prose the probability of E is 0.1031, so that every tenth symbol, roughly, is E. After the 'space' symbol it is the most frequent. For every letter Z occurring in written English the letter E occurs about 200 times. In Table 3 note that the commonest word, 'the', is nine times as frequent as the twentieth most common word, 'at'.

When members of a source S are not equally likely, H(S) is defined as the **average information per member.**

To see how unequal frequencies affect the theory, consider four similar situations in each of which a member of a source is generated by a spinner.

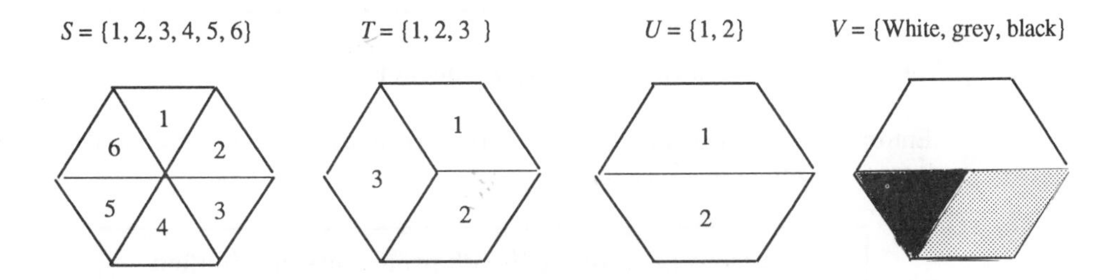

You know that H(S) = $\log_2 6$. The theoretical amount of information per member of S is therefore $\log_2 6$ bits. Similarly, H(T) = $\log_2 3$ and H(U) = $\log_2 2$.

The three members of V have different probabilities and also yield different amounts of information.

'Black' corresponds precisely to '5' of S and so you you would expect it to have precisely the same information content, i.e. $\log_2 6$ bits. Similarly, 'Grey' corresponds to '2' of T and so has information content $\log_2 3$. 'White' corresponds to '1' of U and so has information content $\log_2 2$. The probabilities of Black, Grey and White are $\frac{1}{6}$, $\frac{1}{3}$ and $\frac{1}{2}$ respectively.

The average information per member can be found by multiplying the probability of each member by its information content.

Member	Probability	Information content
White	$\frac{1}{2}$	$\log_2 2$
Grey	$\frac{1}{3}$	$\log_2 3$
Black	$\frac{1}{6}$	$\log_2 6$

If a member of V is chosen at random (by using the spinner) then its average information content is

$$H(V) = \frac{1}{2}\log_2 2 + \frac{1}{3}\log_2 3 + \frac{1}{6}\log_2 6$$

Note that each term is of the form probability x $\log_2\left(\frac{1}{\text{probability}}\right)$

Investigate at least one other source for which you can calculate the information content of each member. Put its average information content into the same form as that given for H(*V*).

The function H is known as the **entropy function.**

'Entropy' is a word borrowed from Thermodynamics, where it is used to denote the degree of disorder in a system.

The entropy of a source *S* with probability distribution $(p_1, p_2, p_3, \ldots, p_n)$ is

$$H(S) = -\sum_{r=1}^{n} p_r \log_2 p_r$$

In calculations of H(*S*) using $\log_2$ you can either use Table 1 or a calculator.

For example, to find $\log_2 0.14$:

Method 1 $\log_2 0.14 = \log_2 \frac{14}{100} = \log_2 14 - \log_2 100$

and Table 1 can be used.

Method 2 $\log_2 0.14 = \dfrac{\log_{10} 0.14}{\log_{10} 2}$

and you can use your calculator.

Check that both methods give the value – 2.837.

Example 1

Find H(*S*) if *S* is a source with six elements having probability distribution

0.05, 0.10, 0.15, 0.20, 0.20, 0.30.

Solution

$$\begin{aligned} H(S) &= -0.05 \log_2 0.05 - 0.10 \log_2 0.10 - 0.15 \log_2 0.15 \\ &\quad -0.20 \log_2 0.20 - 0.20 \log_2 0.20 - 0.30 \log_2 0.30 \\ &= 2.41 \text{ (2 d.p.)} \end{aligned}$$

Check the above calculation, using either of the recommended methods for calculating logarithms.

For a source *S* with members having unequal frequencies, H(*S*) is a theoretical minimum for the average number of questions needed to determine a particular member. This minimum can be attained if you can halve probabilities exactly at each stage. Suppose you have a four-letter source *S* = {A, B, C, D} with probabilities as given in this table.

	A	B	C	D
probability	$\frac{1}{2}$	$\frac{1}{4}$	$\frac{1}{8}$	$\frac{1}{8}$

If you were 'guessing' a member of this source, a good decision tree would be:

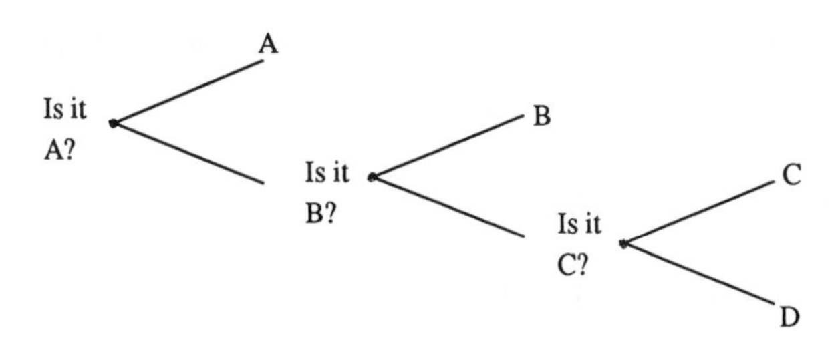

The strategy used is to halve the remaining probability at each stage. Thus at the outset A is just as probable as B, C and D combined. Then if A is not right, B is as probable as C and D taken together.

If you used the tree to identify a typical letter, the average number of questions asked would be

$$(\tfrac{1}{2} \times 1) + (\tfrac{1}{4} \times 2) + (\tfrac{1}{8} \times 3) + (\tfrac{1}{8} \times 3).$$

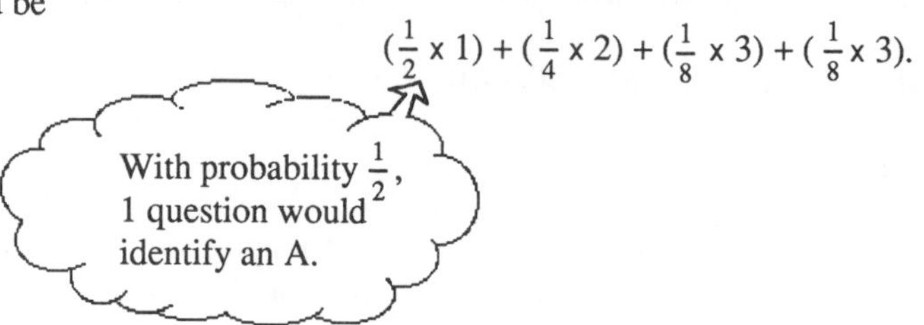

Explain the other terms in the expression.

The average number of questions per letter is therefore 1.75. Note that

$$H(S) = -\tfrac{1}{2} \times \log_2 \tfrac{1}{2} - \tfrac{1}{4} \log_2 \tfrac{1}{4} - \tfrac{1}{8} \log_2 \tfrac{1}{8} - \tfrac{1}{8} \log_2 \tfrac{1}{8}$$

is also 1.75.

Exercise 1

1. Calculate the average information content per letter for this source.

character	A	B	C	D	E
probability	$\frac{1}{2}$	$\frac{1}{4}$	$\frac{1}{8}$	$\frac{1}{16}$	$\frac{1}{16}$

2. Find the entropies of the 4-character sources with probability distributions

   (a) 0.1, 0.2, 0.3, 0.4

   (b) 0.15, 0.20, 0.30, 0.35

   (c) 0.22, 0.24, 0.26, 0.28

   What probability distribution gives the greatest entropy? The connection with the second law of Thermodynamics could be discussed with your Physics teacher.

2.2 Block codes and instantaneous codes

A block code is a code in which all codenames have the same length.

A block code has the advantage that it can be decoded using unsophisticated software; all that has to be done is to lop off the same number of digits every time and read the meaning of the group from a dictionary file. So why bother with other kinds of code?

To answer this, note that the shortest block code for the source

	A	B	C	D
probability	$\frac{1}{2}$	$\frac{1}{4}$	$\frac{1}{8}$	$\frac{1}{8}$

has codenames 11, 10, 01, 00 - all of length 2 digits.

For the same source, the decision tree in Section 2.1 gives the 'tree code'

A	B	C	D
1	01	001	000

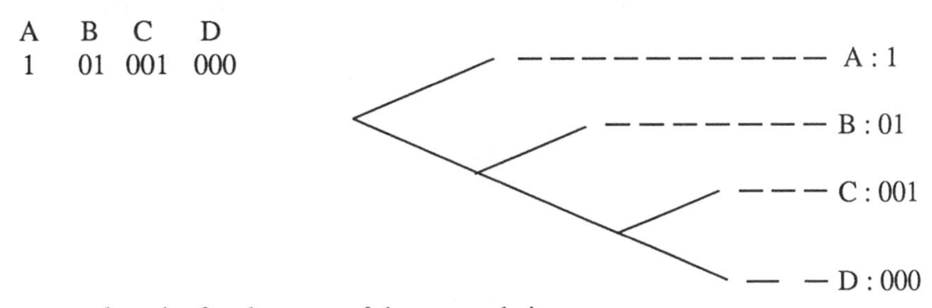

The average length of codenames of the tree code is

$$(\tfrac{1}{2} \times 1) + (\tfrac{1}{4} \times 2) + (\tfrac{1}{8} \times 3) + (\tfrac{1}{8} \times 3) = 1\tfrac{3}{4}.$$

When you use this code rather than the block code fewer digits will encode the same signal. The advantage is most marked when the encoded symbols have very different frequencies. Programs have been written to scan a text, design a code based on word frequency and store both text and decoding algorithm. In some cases this reduces the storage space needed to less than half the original amount.

The generalisation of the expression for average length of codename is straightforward.

If codenames with probabilities $p_1, p_2, \ldots, p_n$ have lengths $l_1, l_2 \ldots l_n$ respectively then the average length of codename is

$$\overline{L} = p_1 l_1 + p_2 l_2 + \ldots p_n l_n = \sum_{r=1}^{n} p_r \, l_r$$

Example 2

Find the average length of codename if the source

	A	B	C	D	E
probability	0.30	0.15	0.05	0.25	0.25

is encoded:

A – 001 B – 01 C – 0001 D – 1 E – 0000

Solution

The average length (in bits) is

$$\overline{L} = (0.30 \times 3) + (0.15 \times 2) + (0.05 \times 4) + (0.25 \times 1) + (0.25 \times 4)$$

$$= 2.65.$$

Do you think the suggested code is well chosen?

TASKSHEET 1 - *Reading codenames*

A code which allows strings of digits to be decoded without ambiguity is said to be instantaneous.

Instantaneous codes are the only ones of practical value. As you will see in question 3 of Exercise 2, all tree codes are instantaneous.

Exercise 2

1. Design an instantaneous code from this tree. Then

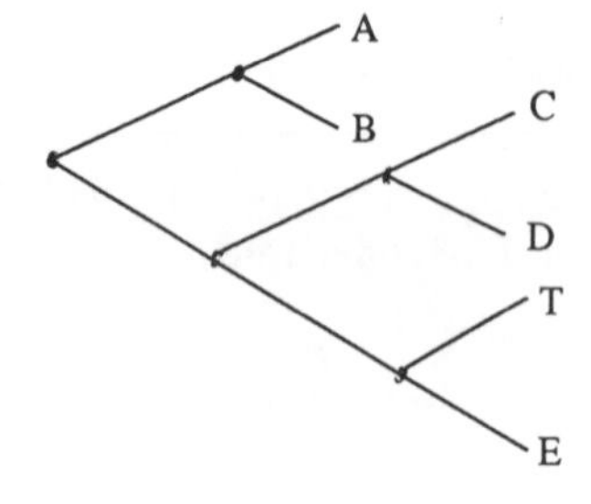

(a) decode the strings of digits

(i) 1011001 (ii) 10000010;

(b) encode the messages

(i) CAB (ii) DEBT

2. $S = \{A, B, C, D, E\}$ with probabilities

A	B	C	D	E
$\frac{1}{4}$	$\frac{1}{8}$	$\frac{1}{4}$	$\frac{1}{4}$	$\frac{1}{8}$

(a) Find a tree code for the source using the 'halving probabilities' principle.

(b) Find H(S).

(c) Suppose that you encode the same source S, with the same probability distribution, as follows:

A	B	C	D	E
1	0001	01	001	0000

What would be the average length of codename?

(d) What do you deduce from answers to (b) and (c) above?

3. (a) How can you quickly spot that

A – 1   B – 01   C – 10   D – 00

is not an instantaneous code? Deduce a general test to check that a code is instantaneous.

(b) Show that all tree codes are instantaneous.

4. $S = \{A, B, C, D, E\}$ with probabilities

A	B	C	D	E
0.37	0.22	0.18	0.15	0.08

(a) Find H(S).

(b) Find the best tree code you can for this source.

(c) Find the average length of a codename in your tree code.

(d) Guess a general rule from answers to (a) and (c).

2.3 Compact codes

You have seen that for sources whose members have very different probabilities, by using codenames of unequal length rather than a block code, you achieve a significant reduction in the number of bits used.

A compact code is an instantaneous code for which $\bar{L}$ is as low as possible.

The least possible average length of codename cannot be less than the entropy of the source, i.e. $\bar{L} \geq H(S)$. A proof of this result is beyond the scope of this unit.

For the source

	A	B	C	D
probability	$\frac{1}{2}$	$\frac{1}{4}$	$\frac{1}{8}$	$\frac{1}{8}$
codename	1	01	001	000

you found that the uncertainty was 1.75 and the average length of codename was also 1.75; the code was as efficient as possible. In such a case the efficiency is 100% . In general

The efficiency of a source encoding is defined as

$$\text{efficiency} = \frac{H(S)}{\bar{L}}$$

Example 3

A source with probability distribution (0.05, 0.10, 0.15, 0.20, 0.20, 0.30) has entropy 2.41. Find the efficiency of the encoding: 0000, 0001, 001, 01, 10, 11.

Solution

$$\bar{L} = (0.05 \times 4) + (0.10 \times 4) + (0.15 \times 3) + (0.20 \times 2) + (0.20 \times 2) + (0.30 \times 2)$$

$$= 2.45$$

The efficiency of this encoding is

$$\frac{2.41}{2.45} = 98.4\%.$$

The following tasksheet describes the two main methods used to obtain compact binary codes. The fact that both methods yield compact codes may be assumed without proof.

TASKSHEET 2 - *Efficient binary codes*

2.4 A note on the psychology of alphabets

Most computers consist of two-state devices, either 'off' or 'on', and so the binary alphabet (0, 1) is preferred by them. On the other hand people prefer to discriminate among more objects than two. The world generally has adopted the ten decimal digits and most languages use alphabets with between 16 and 36 letters, each of which may be written in upper or lower case.

For this reason, computer operators often use alphabets which are based on binary representation but have more characters than the binary alphabet. The most common are the **octal** and the **hexadecimal** systems. In octal numbers the base or **radix** is 8 and in hexadecimals it is 16. Thus

$$(83)_{10} = (1 \times 64) + (2 \times 8) + 3 = (123)_8$$

and

$$(83)_{10} = (5 \times 16) + 3 = (53)_{16}.$$

So the octal form of the decimal number 83 is 123 and the hexadecimal form is 53.

Just as for decimal numbers you need 10 digits, for hexadecimal numbers you need 16: 0 to 9 and then extra digits standing for 10, 11, 12, 13, 14 and 15. These extra digits are usually taken to be upper case letters, as follows:

decimal	0 1 ... 9	10	11	12	13	14	15
hexadecimal	0 1 ... 9	A	B	C	D	E	F

Example 4

Convert the hexadecimal number D7 into binary form.

Solution

$$D7_{16} = (13 \times 16 + 7)_{10}$$

$$= (1101 \times 10000 + 111)_2$$

$$= 11010111_2$$

Find another method of converting $(D7)_{16}$ to binary form. Find a way of converting $(D7)_{16}$ to octal form.

Exercise 3

1. Explain how the octal and hexadecimal systems are based on binary representation. As an example, consider the binary, octal and hexadecimal forms of the decimal number 37.

2. Justify the witticism that Christmas is the same as Halloween because 25 Dec is 31 Oct.

3. What is the binary representation of the hexadecimal E8?

4. Make a multiplication table for octal numbers (up to 7 x 7).

After working through this chapter you should:

1. understand and use the terms **entropy** of a source and **block, instantaneous** and **compact**, as applied to codes;

2. be able to calculate the entropy of a given finite source and the efficiency of an encoding of that source;

3. be able to use the methods of Fano and Huffman to devise compact codes;

4. appreciate that the entropy of a source is a lower bound for the average length of codenames in any encoding of that source;

5. be able to use binary, octal and hexadecimal representations of numbers in calculations;

6. be familiar with the ASCII code.

Reading codenames

SOURCE ENCODING

TASKSHEET 1

You already have two encodings for the source {A, B, C, D}.

Code 1 (block code)

A – 11, B – 10, C – 01, D – 00

Code 2 (names of unequal length)

A – 11, B – 01, C – 001, D – 000

1. Decode these strings in code 1.

   (a) 101100 (b) 0001101111.

2. Decode these strings in code 2.

   (a) 0011101 (b) 10001110001.

Another possible code is as follows:

Code 3

A – 1, B – 01, C – 10, D – 00

3. Decode these strings in code 3.

   (a) 101001 (b) 100011

4. Write out your own strings of binary digits (randomly, using say, the toss of a coin to decide whether 1 or 0 comes next). Decode them using codes 1, 2 and 3.

5. Which of the codes are of practical value and which not? Explain your answer.

Efficient binary codes

SOURCE ENCODING

TASKSHEET 2

Two main methods are used to devise very efficient binary codes. They are named after Fano and Huffman, who were early workers in coding theory.

Fano method

The Fano method is easily illustrated by a decision tree and consists essentially of repeated division into equally likely groups.

To illustrate the method consider the source S:

Letter	A	B	C	D	E	F	G	H
Probability	0.12	0.21	0.08	0.10	0.04	0.15	0.25	0.05

The elements are first listed in descending order of probability.

G	B	F	A	D	C	H	E
0.25	0.21	0.15	0.12	0.10	0.08	0.05	0.04

A decision tree is then formed by halving as nearly as possible the probabilities at each stage. Since the total probability is 1 the first step is to run down the list until you have a total probability as near as possible to 0.5. This divides the list into

GB FADCHE

The complete process is shown below.

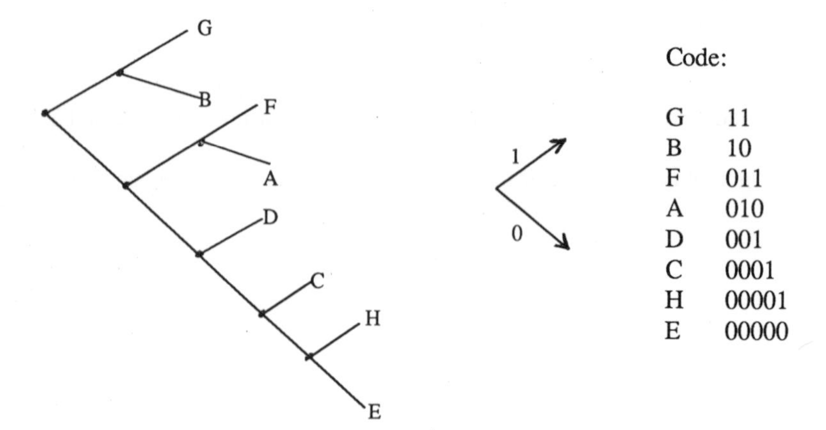

(continued)

Huffman's method

Huffman's method is rather more difficult to describe, though easy to apply. To illustrate it, use the same source, *S*.

Letter	A	B	C	D	E	F	G	H
Probability	0.12	0.21	0.08	0.10	0.04	0.15	0.25	0.05

Stage 1 Rearrange, if necessary, in descending order of probability.

G	B	F	A	D	C	H	E
0.25	0.21	0.15	0.12	0.10	0.08	0.05	0.04

Stage 2 Identify the two least probable letters (H and E) and treat (H, E) as a single entity, with probability 0.05 + 0.04 = 0.09. Then reorder as in Stage 1.

G B F A D (H,E) C

Now treat (H,E,C) as a single entity as in Stage 2 and repeat the process until you are left with only two groups. You should then have these successive listings.

G	B	F	A	D	C	H	E
	G	B	F	A	D	(H,E)	C
		G	B	(H,E,C)	F	A	D
			G	(A,D)	B	(H,E,C)	F
				(H,E,C,F)	G	(A,D)	B
					(A,D, B)	(H,E,C,F)	G
						(H,E,C,F,G)	(A,D,B)
						↑	↑
						0	1

To illustrate the encoding procedure, take as an example the letter A. Look for the listings in which it occurs in one of the last two positions, starting from the bottom and working to the top. The codename of A is 100 because A occurs in the last position (row 7), in the next to last position (row 5) and then the next to last position (row 3).

(continued)

You should check your understanding of the encoding procedure by finding the rest of the codenames before reading on. They are as follows:

A	B	C	D	E	F	G	H
100	11	0001	101	00001	001	01	00000

The average length of codename is

$$\bar{L} = (0.12 \times 3) + (0.21 \times 2) + (0.08 \times 4) + (0.10 \times 3) + (0.04 \times 5) + (0.15 \times 3) + (0.25 \times 2) + (0.05 \times 5)$$

$$= 2.8.$$

The average information per letter of the source (the entropy)

$$H(S) = -(0.12 \log_2 0.12 + 0.21 \log_2 0.21 + \ldots + 0.05 \log_2 0.05)$$

$$= 2.776 \text{ (to 3 d.p.).}$$

The efficiency of the code is $\dfrac{H(S)}{\bar{L}} = 99.1\%$.

You will notice that, though the digits in the codenames may differ, the Fano and Huffman procedures yield the same distribution of lengths of codename and so the same efficiency. Both give compact codes for the source S. In practice, Huffman's method is more often used.

Questions

1. Find two distinct Huffman codes for the four-letter sources with probability distribution $(\frac{1}{6}, \frac{1}{6}, \frac{1}{3}, \frac{1}{3})$ and show that they result in equally efficient codes.

2. Use Fano's procedure to encode these sources. Find the efficiency of your code in each case.

   (a)

Letter	A	B	C	D	E
Probability	0.12	0.61	0.15	0.04	0.08

   (b)

Letter	A	B	C	D	E	F	G
Probability	0.12	0.38	0.09	0.25	0.08	0.03	0.05

3. Using Table 2, show that the average length of Huffman codenames for the English alphabet with space is less than the length of codenames in a block code for that source.

4. (a) For which sources is the Huffman code simply a block code?

   (b) For which sources does Huffman encoding give a significant economy over the block code?

5. For which sources does the Huffman procedure give codes with 100% efficiency?

The ASCII code

SOURCE ENCODING

TASKSHEET 3

Many computers use the seven-bit ASCII code to give a source alphabet of $2^7 = 128$ characters.

Seven-Bit ASCII Code

Octal Code	Char-acter	Octal Code	Char-acter	Octal Code	Char-acter	Octal Code	Char-acter
000	NUL	040	SP	100	@	140	`
001	SOH	041	!	101	A	141	a
002	STX	042	"	102	B	142	b
003	ETX	043	#	103	C	143	c
004	EOT	044	$	104	D	144	d
005	ENQ	045	%	105	E	145	e
006	ACK	046	&	106	F	146	f
007	BEL	047	'	107	G	147	g
010	BS	050	(	110	H	150	h
011	HT	051	)	111	I	151	i
012	LF	052	*	112	J	152	j
013	VT	053	+	113	K	153	k
014	FF	054	,	114	L	154	l
015	CR	055	—	115	M	155	m
016	SO	056	-	116	N	156	n
017	SI	057	/	117	O	157	o
020	DLE	060	0	120	P	160	p
021	DC1	061	1	121	Q	161	q
022	DC2	062	2	122	R	162	r
023	DC3	063	3	123	S	163	s
024	DC4	064	4	124	T	164	t
025	NAK	065	5	125	U	165	u
026	SYN	066	6	126	V	166	v
027	ETB	067	7	127	W	167	w
030	CAN	070	8	130	X	170	x
031	EM	071	9	131	Y	171	y
032	SUB	072	:	132	Z	172	z
033	ESC	073	;	133	[	173	{
034	FS	074	<	134	\	174	|
035	GS	075	=	135	]	175	}
036	RS	076	>	136	^	176	~
037	US	077	?	137	—	177	DEL

In the table each character is shown with its octal representation. Note that only the second and third octal digits should be expanded to give three binary digits when converting to binary form.

Example:

character	octal code	binary code
n	156	1101110

(continued)

Since computers usually work in 8-bit **bytes** each character is normally given an eighth binary digit, which may be used either

- to provide an even parity check (see Error detection in Chapter 4)

or

- as a timing device, in which case it is always 1.

Questions

1. Give the binary codenames for the characters R and r.

2. What meaning would you give to the binary codename

   11 111 000?

3. Give the codenames of the decimal digits 0-9 in binary form and make a comment on them.

4. Encode the lower-case alphabet using ASCII code

   (a) base 4 (b) base 16

5. The instruction for converting 7-bit ASCII to the LT 33 8-bit teletype code is

   $$\text{LT } 33 = \text{ASCII} + (200)_8.$$

   Give the octal and binary representations of the letter R in LT 33. Give also an alternative answer to question 2.

Tutorial sheet

SOURCE ENCODING

TASKSHEET 4

1. A letter is chosen at random from a bag of Scrabble letters and you are told whether it is a consonant, vowel or blank. What is the average information of this knowledge?

2. (a) Calculate the information content of the reference number

   JEN/05271/09/82

   assuming that all letters are used with equal frequency and all decimal digits are used with equal frequency.

   (b) Now suppose that JEN is the first part of a surname and that 09/82 is a date, referencing having started in January 1981. Stating clearly the assumptions made, estimate the information content of the same reference number.

3. Explain why, in a compact code, the two least frequent source members always have codenames of the same length.

4. A worker in Coding Theory named Zipf pointed out that in many sources the frequency of the kth most probable item is roughly proportional to $\frac{1}{k}$. For this reason, a model in which probabilities are in the ratios $1 : \frac{1}{2} : \frac{1}{3} : \frac{1}{4} : \ldots$ is called a Zipf model.

   (a) Comment on how well the alphabet (with space) and the word-list in Tables 2 and 3 fit Zipf's model distribution.

   (b) Find the entropy of a source of 10 items having Zipf's probability distribution.

   (c) Find the Huffman code for the source in (b) and calculate its efficiency.

3 *Redundancy*

3.1 What is redundancy?

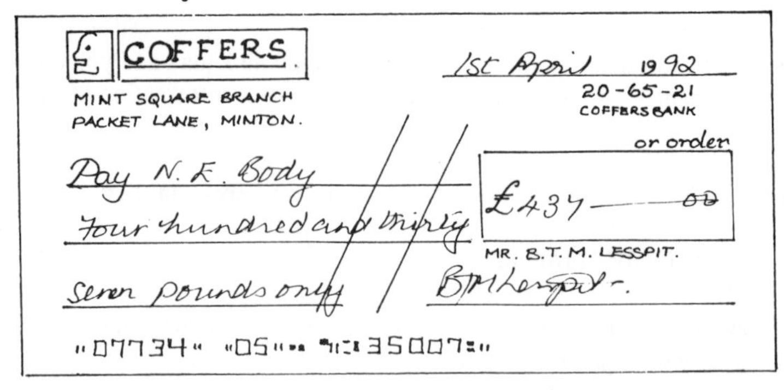

On a cheque the amount is written twice, in case some figure or word is difficult to decipher. Similarly, when radio broadcasters give a telephone number over the air it is normally repeated. If you get the message first time its repetition is useless. However, because of the 'noise' in the communication, caused by bad writing or inattention, it is a valuable safeguard to introduce what appears to be extraneous information. This extraneous information is known as **redundancy**.

Sometimes, for example when listening to a formal lecture, you feel that the stream of information your brain is being asked to process is more than you can take and you long for a bit of light relief: a funny story or even just a review of the main points of the lecture so far. In fact, lecturers cannot avoid a great deal of redundancy because it is inherent in the language.

TASKSHEET 1 - *Clarity and brevity*

The English language contains a great deal of redundancy, which is essential to its users.

In fact, as will be seen, in all practical forms of communication redundancy is deliberately introduced in order to combat noise.

Identify instances of redundancy and the noise it is intended to overcome in:

1. **political speeches;**
2. **road signs;**
3 **shop fronts;**
4. **signs on public lavatories at airports.**

3.2 Measuring redundancy

To show how redundancy is measured, the English language may be taken as an example. For simplicity, only the 26 letters and the space will be used, omitting punctuation marks. If these 27 symbols were equally likely, the average information content per symbol would be 4.755.

Why? How would you obtain such a source?

$\log_2 27 \approx 4.755$ is sometimes written $H_0(S)$ and a random stream of the 27 symbols with equal frequency is called the **zeroth approximation** to English. As the name implies, it is not really an approximation to English at all, since it does not even take into account the frequencies with which the letters are used.

A better model uses these frequencies, as given in Table 2. Recalling that the average information for a source with probability distribution $(p_1, p_2 \ldots, p_n)$ is

$$-\sum_{r=1}^{n} p_r \log p_r$$

you can calculate the average information in a **first approximation** to English. This average information is called $H_1(S)$.

Calculate $H_1(S)$.

A source has the greatest average information per member when the members of the source are equally likely. It follows that in the English language the 27 symbols would be used to best effect (from the viewpoint of an information theorist!) if they were equally frequent. This would be a zeroth approximation, with information content 4.755 bits per member. Any other approximation is compared to this to measure its redundancy. In effect, redundancy is defined as the ratio of 'wasted information' to maximum possible information.

$$\text{Redundancy} = \frac{\text{wasted information}}{\text{maximum possible information}}$$

For the first approximation to English

$$\text{wasted information} = 4.755 - \text{average information per symbol}$$

$$= 4.755 - H_1(S)$$

$$\text{redundancy} = \frac{4.755 - H_1(S)}{4.755}$$

In general:

For any source S with n members,

$$\textbf{redundancy} = \frac{\log_2 n - H(S)}{\log_2 n}$$

Exercise 1

1. Find the redundancy of the source

   (a)

A	B	C	D
$\frac{1}{2}$	$\frac{1}{4}$	$\frac{1}{8}$	$\frac{1}{8}$

   (b)

A	B	C	D	E
0.20	0.40	0.30	0.05	0.05

2. Find the redundancy of the first approximation to English.

3. Find the efficiency of the Huffman code for the alphabet with space, regarded simply as a set of symbols with the probabilities given in Table 2.

Possible project

Estimating the redundancy of the English language is a good subject for an extended investigation.

One method of finding the average information content per symbol is as follows. Ask a partner to read out half a page of a novel. This should give you an idea of the story and the author's style. Take up the story at the point at which your partner left it and ask yes/no questions to construct a further passage of 100 symbols, keeping a tally of the number of questions asked. You should use your knowledge of spelling and syntax, of the sense and style of the language and also of the frequencies of the letters. (Table 2 will be helpful in your questioning; you can show the list and ask a question like: is it above R in this table?) When you are sure of a symbol then there is no need for a question and you can simply write down the symbol without adding to the tally.

True English transcends any approximation, so the average information you find is sometimes represented by the symbol $H_\infty(S)$.

3.3 Digrams

A digram is an ordered pair of symbols. In Table 4 you can see a frequency count for the most common digrams in the English language, taken from a sample size of 10^4.

> **(a) With 27 symbols, how many possible digrams are there?**
>
> **(b) Why do you not need to consider all of them?**

If you take into account the digram structure you can make a **second approximation** to the English language. The average information per symbol in this case is denoted by $H_2(S)$.

In theory there is no end to the sequence of approximations; you could make a table of probabilities of the 27^3 **trigrams** and use it to generate a **third approximation**, and so on. In practice the task of constructing such tables becomes too great, even with computer assistance, after the digram case. A useful table of probabilities of digrams can be constructed from a single book, whereas a table of trigram probabilities would require a small library.

Possible project

Construct a sample of a second approximation to English, using an 'interlocking digrams' method, as follows. Open a novel fairly near the beginning and select a digram at random somewhere in the middle of the page, say RT. Read on from this point until you come to the next occurrence of the second element of the digram (T in this example) and note the following symbol, so you now have, say, RT □ . Continue reading until you come to the next occurrence of the latest symbol noted and write down the symbol following that in the text, and so on. So your sample might start: RT □ PEN □ ... from a second approximation you can often identify a language. You might like to use the same method to obtain a second approximation to French or Latin, say, for comparison.

> After working through this chapter you should:
>
> 1. appreciate the need for redundancy in communication, particularly when using spoken or written language;
> 2. be able to calculate the redundancy of a given source;
> 3. understand the meaning of an *r*th approximation to a language.

Clarity and brevity

REDUNDANCY

TASKSHEET 1

1. Try to reconstruct this passage, from which the small words have been omitted.

   When binary codename transmitted communication channel, possible signal affected noise, usually effect corrupting random selection transmitted bits, 1 received 0 vice versa.

2. Now try this one.

   Whn bnry cdnm s trnsmttd vr cmmnctn chnnl t s pssbl tht th sgnl wll b ffctd by ns. Ths wll slly hv th ffct f crrptng rndm slctn f th trnsmttd bts, s th l s rcvd s 0 nd vc vrs.

You probably spotted that these are transcriptions of the same piece of English. In passage 2 the vowels are missing. (In written Hebrew, unless it is written for children, the vowels are normally omitted.) You probably could have reconstituted the original piece from either of these passages, at least sufficiently well to understand it; though someone unfamiliar with the subject matter might be puzzled by the 'ns' in the second passage. However, if a lecturer inflicted version 1 on you, or you were expected to read version 2, you would soon become irritated and would probably stop trying to follow.

3. Write the passage as you might say it over the telephone, including any redundancy which might be needed to overcome noise.

Tutorial sheet

1. (a) In a table of digrams used in constructing a second aproximation to English, list four digrams (not having Q as first letter) all of which would have probability zero.

   (b) Given that the entropy of the second approximation $H_2(S) = 3.32$, calculate the redundancy of the digram source.

2. Consider a two-member source with probability distribution

   $(p, 1-p)$

   (a) Write down an expression for the entropy $H(p)$ of the source.

   (b) **Either** draw a graph of $H(p)$ against p **or** differentiate $H(p)$ with respect to p. Hence show that $H(p)$ takes its maximum value when $p = \frac{1}{2}$.

3. The result of question 2 is a particular case of the theorem:

> **For a finite source *S*, H(*S*) is greatest when the members of *S* have equal probability.**

Assuming this theorem, how do you think teachers should allocate grades such as A, B, C, D and E when marking?

Possible project

Find a convincing demonstration of the theorem quoted above when the number of members of S is greater than 2.

4 *Channel encoding*

4.1 Error protection

The main problem of coding theory is that of striking a balance between two opposing needs. It could be said that there are two kinds of encoding. In **source encoding** the object is to use the code source (usually (0, 1)) as efficiently as possible, subject to the limitation that the entropy of the source imposes a lower bound on the average code length.

Express this last condition in your own words.

In **channel encoding** the objective is to make sure that codenames are easily distinguished, so that a small change due to noisy transmission does not change one codename into another. The compromise made for the encoding will depend on the noisiness of the channel and the importance of accurate transmission.

Discuss the levels of error protection in various forms of written English. Why is there a tendency to introduce more redundancy in spoken than in written English?

You have seen that when encoding thoughts into words some redundancy must be introduced in order to overcome noise. The same applies to other codes in practical use.

To take an extreme case, transmissions from the Mariner-9 spacecraft of encoded pictures of Mars were packaged into 6-bit groups each of which was sent with 26 bits of protection as a 32-bit string.

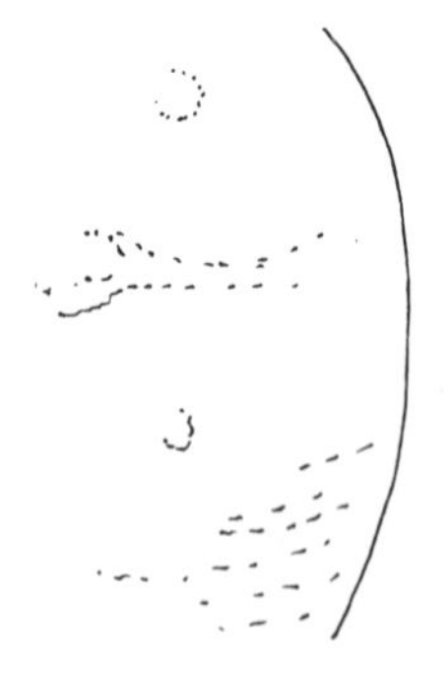

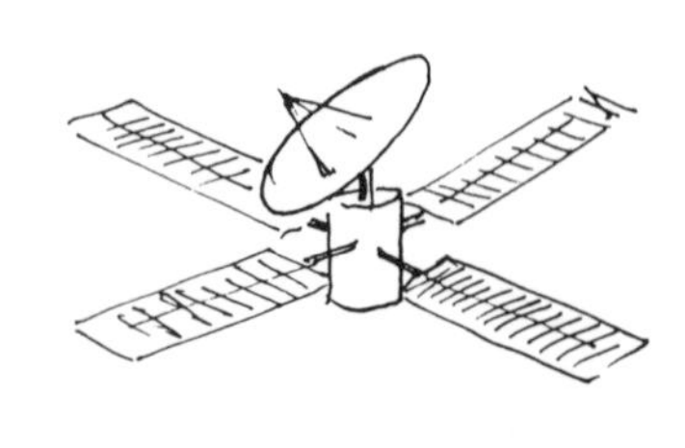

This seems a huge amount of redundancy but it achieved a near miracle: a tiny 20-watt transmitter was able to send, over 84 million miles of space, pictures which were of high enough quality to distinguish fine detail on the surface of Mars. This can be contrasted with the Crystal Palace transmitter which has a power of 40 000 watts and a range of only 60 miles.

Before examining the Mariner-9 code in detail, it is necessary to establish the principles of error protection, in which redundancy is deliberately introduced into signals to combat noise.

4.2 Error detection

The simplest form of error protection involves the detection of an error in a signal. When there is two-way communction between sender and receiver this may be adequate; if an error is spotted the receiver can request a repeat of the signal, which should put things right.

The most obvious method of protection is that used in writing a cheque: namely repetition. In its simplest form, each digit is repeated as many times as necessary, giving R2, the two-fold repetition code, R3 the three-fold code and so on.

	Message	Codename		Message	Codename
R2	0	00	R3	0	000
	1	11		1	111

If R2 is being used and 10 is received then an error is detected and the receiver asks for a repeat.

In a more sophisticated method of error detection, every codename has one or more **check digits** in addition to each group of message digits. For example, the English language can be encoded using a block code with five-digit groups. Such a code is the International Telegraph Alphabet (ITA), given in Table 5. To show up errors another binary digit could be appended to each group so that the total number of 1's in the final codename is even. This device is called a **parity check** and the check digit is called a **parity digit.** For example, the ITA group for letter A is 00111. To form its protected codename you should append the check digit 1

(a) Give the ITA codenames with a check digit giving even parity for (i) B (ii) ? (iii) 7.

(b) Decode the protected codenames

(i) 100010 (ii) 001001 101000 (iii) 111010.

(c) What action would you take if you received the signal 111110?

(d) Under what circumstances would this form of error protection fail?

4.3 Efficiency and redundancy

You have already met the idea of efficiency of a source encoding. This sections deals with channel encoding efficiency which depends upon the number of bits used for protection.

> **The *efficiency* of a code is the $\dfrac{\text{number of message bits}}{\text{total length of codename}}$.**
>
> ***Redundancy* is defined to be 1 – efficiency.**

So the protected ITA code has efficiency $\frac{5}{6}$ and redundancy $\frac{1}{6}$.

> **What are the efficiency and redundancy of the repetition code R2 ?**

A useful notation for block codes is the **(n, k)** form as defined below:

> **Block codes in which the codenames are of length n and have k message digits are called (n, k) codes .**

> **(a) Explain why the protected ITA code is a (6, 5) code.**
>
> **(b) Classify the repetition code R3 in (n, k) form.**

Exercise 1

1. (a) What is the efficiency of an (n, k) code?

   (b) Describe the s-fold repetition code Rs using the (n, k) notation.

2. A student sometimes communicates for fun with her younger brother using the form of protected ITA described above. She is not always as careful as she might be with her transcription. If her brother receives the signal

   010111 011110 110000 011011 110110 010111 111001 100011 000011 010111

   what should he make of it?

3. (a) Write out the complete (4, 4) code. What is its efficiency?

   (b) How many elements has the complete (n, n) code?

4. (a) In numerical work you might make use of the **2-in-5 code**, whose codenames all have five digits of which exactly two are 1's.

Thus 10001 is included but 11100 is not.

List the full code and explain why it is useful for numerical data. What do you think is the efficiency of the code?

(b) A similar code, used in radio telegraphy, is the **3-in-7 code** (the **van Duuren code**). How many codenames are there? What is the efficiency of the code?

You should be familiar with each of the following codes used for channel encoding.

R*n*	**The *n*-fold repetition code in which each bit is repeated *n* times.**
Even-weight code of length *n*.	**The block code with *n*-digit codenames such that the total number of 1's in each codename is even.**
ITA	**The International Telegraph Alphabet.**
2-in-5 code	**The block code with 5 digit codenames such that the number of 1's in each codename is precisely 2.**

4.4 Error correction

Telephone companies use binary signals for transmission of data to FAX machines and microprocessors generally. Such signals are transmitted by two-state electrical systems as square waveforms in **pulses.**

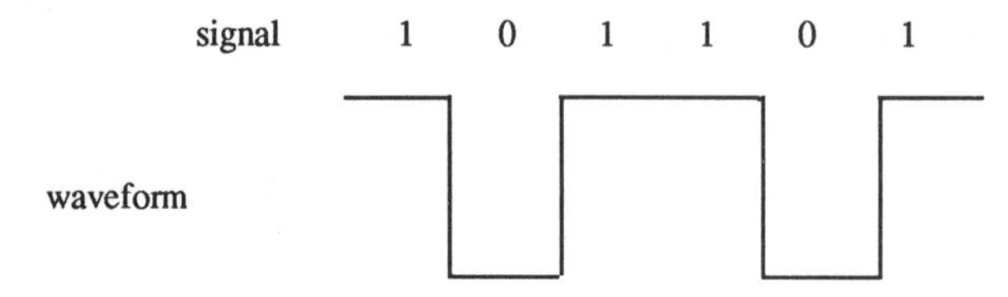

Corruption of binary signals is most often caused by distortion of the waveforms. After travelling through a mile of cable the waveform above might look like this.

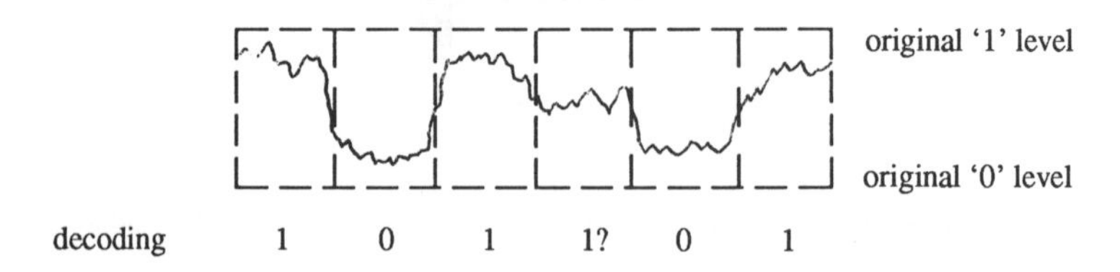

In practice, companies such as British Telecom send binary signals with protection and in addition relays called pulse regenerators are introduced into the line at regular intervals to reconstitute the uncorrupted signal.

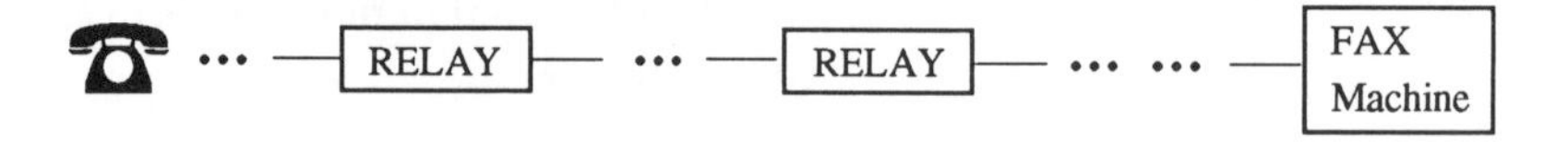

When it can be accomplished easily , and certainly when the communication is one-way only, corrupted binary signal groups should not just be detected; the faulty digits should be identified and corrected. One of the pioneers in applying mathematics to the communication of information was C.F. Shannon. He showed that, by using sufficient extra digits, information could be transmitted to any desired degree of accuracy through a noisy channel. The designers of the Mariner-9 system were no doubt fortified by that knowledge.

The simplest method of error correction is known as **majority logic decoding**. The same signal is transmitted many times with the assumption that the one most often received is probably right. The crudest way of using the method employs the repetition codes. Thus if the signal 101 were sent using 5-fold repetition, through a noisy channel, the received signal might be

10111 01100 11111

Though this would be accurately decoded, the second group of bits indicates that 5-fold repetition is barely adequate and the use of a greater number of repetitions might need to be considered.

Compare the effect of a single error in a group using the first few repetition codes.

		Message	Code	With error	Diagnosis/action
(a)	using R2	1	11	10	error detected
(b)	using R3	1	111	101	error detected and corrected by majority logic decoding
(c)	using R4	1	1111	1011	error detected & corrected by majority logic decoding

Note that if there were two errors they would be detected but not corrected by R4. Using R2, you would not know whether the corrupt signal 10 stood for 1 or 0 and in R4 you would have similar difficulties with 1010. Here is a table similar to that above, in which two errors are introduced.

		Message	Code	With errors	Diagnosis/action
(d)	using R4	1	1111	1010	2 errors detected, not corrected
(e)	using R5	1	11111	10101	2 errors detected and corrected by majority logic decoding

Exercise 2

1. Complete this table, which shows the numbers of errors detected and corrected using the repetition codes Rn (n = 2, 3, ...).

n	number of errors detected	number of errors corrected
2	1	0
3	1	1
4	2	1
5		2
6		
7		

How many errors may be detected and how many corrected using the code Rn ?

2. (a) The even-weight code of length 4 has codenames

0000 1100 1010 1001 0110 0101 0011 1111

What error protection does it give in terms of the numbers of errors which can be detected and corrected?

(b) Write out the codenames of the even-weight code of length 5. What error protection goes it give?

(c) For the even-weight code of length n state

(i) the number of codenames (ii) the error protection.

4.5 Hamming's error-correcting system

R.W. Hamming was the inventor of a system which identifies a single faulty digit in a codename with the most economical possible use of check digits. The method can be illustrated by an example in which three check digits are appended to a four-bit code-name in the first, second and fourth positions.

[x] [x] [ ] [x] [ ] [ ] [ ] ([x] check digit)

The check digits are binary digits calculated so as to make the sums of the digits in the

1st, 3rd, 5th and 7th
2nd, 3rd, 6th and 7th
4th, 5th, 6th and 7th
} positions all even.

These particular combinations of positions are chosen because:

the numbers 1, 3, 5, 7 have binary representation * * 1
the numbers 2, 3, 6, 7 have binary representation * 1 *
the numbers 4, 5, 6, 7 have binary representation 1 * *

For example, the message group 0010 would be protected as follows:

Message group 0010

position	1	2	3	4	5	6	7		
	-	-	0	-	0	1	0	←	message group
	0	1		1				←	check digits calculated
	0	1	0	1	0	1	0	←	protected signal

To see how the check works, suppose the fifth bit is received incorrectly.

0 1 0 1 1 1 0 ← signal with error

The error in the faulty signal is located by carrying out the checks:

Check 1	(positions 1, 3, 5, 7)	0 + 0 + 1 + 0	not even : incorrect
Check 2	(positions 2, 3, 6, 7)	1 + 0 + 1 + 0	even : correct
Check 3	(positions 4, 5, 6, 7)	1 + 1 + 1 + 0	not even : incorrect

Assuming that there is only one error, it must be in the position whose number has binary representation

* * 1	(check 1)
* 0 *	(check 2)
1 * *	(check 3)

The error is therefore in position $101_2 = 5$. The correct signal is 0101010 and the message is 0010.

> **(a) Show that the codename 1111111 is correct.**
>
> **(b) Correct and decode the faulty codename 1001111.**

In general, c check bits can be used to make c checks:

check 1 on bits whose positions are $* \,.\,.\, * * 1_2$
check 2 on bits whose positions are $* \,.\,.\, * 1 *_2$
etc.

The c checks can therefore deal with codenames which have length

$$\underbrace{11\ldots11_2}_{c \text{ bits}} = 2^c - 1.$$

The $2^c - 1$ digits of the codename include c check digits and so the number of message digits is at most

$$2^c - 1 - c.$$

The check digits need not be in the first, second, fourth ... positions in the actual signal. For example, they might be at the end of the codename so that they can be easily lopped off to obtain the message.

Exercise 3

1. Write out the complete Hamming error-correcting code with 2 check digits.

2. Locate and correct the error in the codename

   011100010111110

   given that the check digits are in the first, second, fourth and eighth positions.

3. (a) If all message digits are used, show that the redundancy of the Hamming error-correcting code with c check digits is $\frac{c}{2^c - 1}$.

   (b) In the example with 3 check digits, the redundancy of $\frac{3}{7}$ is very high. How efficient are error-correcting codes with more check digits?

4.6 Continuous signals

So far you have considered only sources which are finite and (hence necessarily) **discrete**; that is, having completely distinct members. Many sources of great importance, however, are **continuous** in nature; in particular, speech and music are most easily rendered into continuous electrical signals, varying in amplitude and frequency to represent the variations in loudness and pitch of the sound.

The effect of noise on two-stage signals is far less than that on systems in which there are an infinity of states because it is easy for the receiver to distinguish between two states electrically and hence to correct minor corruption. In addition, as you have seen, error protection can be applied to a binary signal to correct more serious corruption. A further advantage of binary signals is that they may be transmitted very rapidly.

For these reasons it is advantageous to convert continuous into two-stage systems, and this is done in transmitting speech, music and visual information. In fact all these, together with telephone signals and digital data, may be transmitted in any mixture through the same channel.

When any waveform is analysed, its parts are made up of a set of regular waveforms of various amplitudes and frequencies. The complete set of frequencies for a waveform is called its **frequency spectrum**. The **Sampling Theorem** states that:

> **If a waveform's frequency spectrum contains no frequencies higher than W Hz, then the waveform is completely determined by sampling its values at a set of points less than $\frac{1}{2W}$ seconds apart.**

Hence all waveforms encountered in practice may be represented by a discrete set of points obtained by **sampling** the waveforms at sufficiently small intervals of time.

> **For many purposes a top frequency of 8000 Hz is assumed. How often should the waveform be sampled?**

The technique used to render continuous signals into binary form is called **pulse code modulation (p.c.m.).**

P.c.m. is carried out in stages. In the first, the amplitude is sampled at regular intervals. The sample amplitudes are then 'quantised' - i.e. each given one of a discrete set of equally spaced values - and finally the quantised values are encoded in binary form. The process is illustrated in the following example.

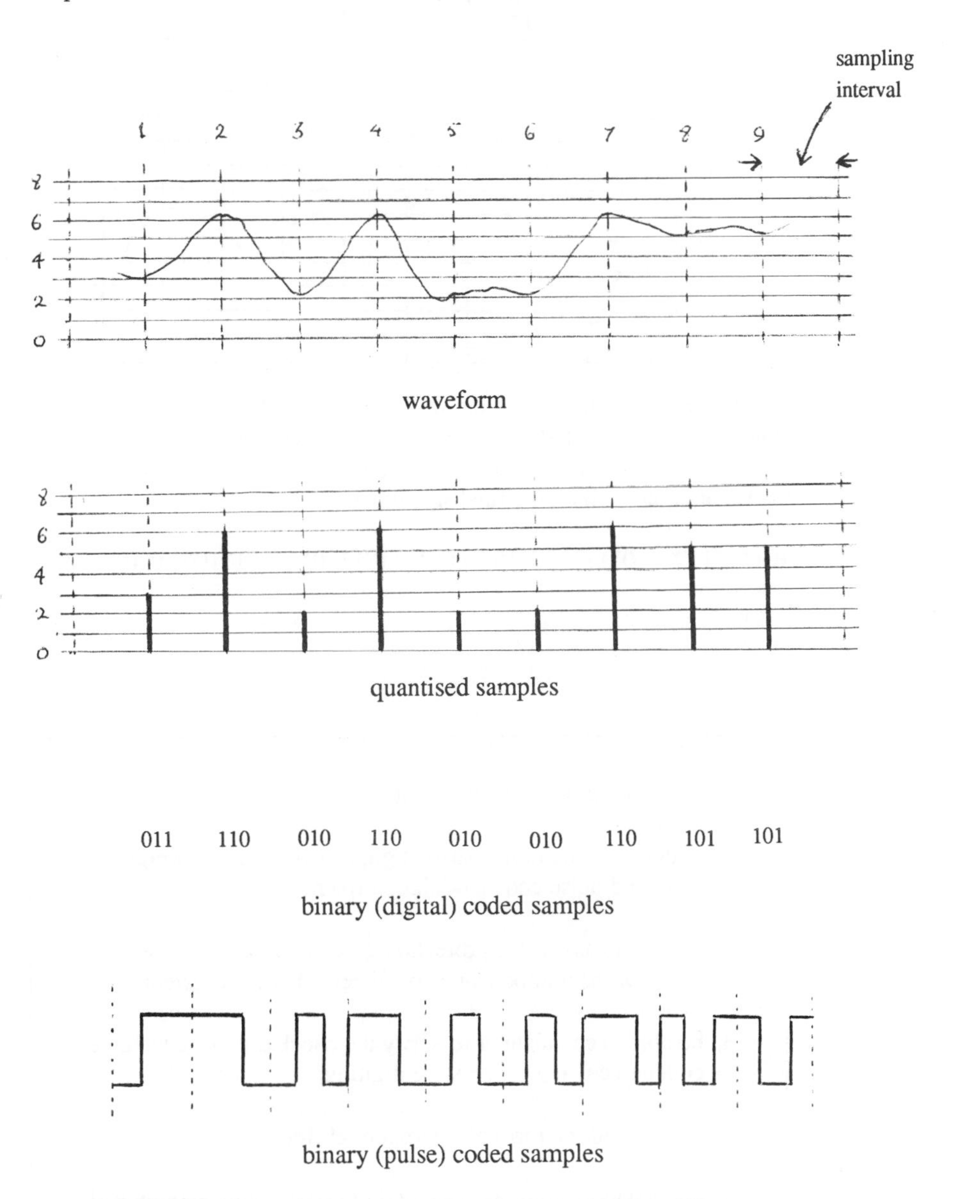

waveform

quantised samples

binary (digital) coded samples

binary (pulse) coded samples

In the example given, the quantisation is into 8 levels, and this is sufficient to render speech intelligible. For music, as in digital recording, at least 16 levels are needed and 32 are commonly used.

Exercise 4

1. Use p.c.m. to find the binary code for this waveform, using the 16 levels at the 10 times indicated.

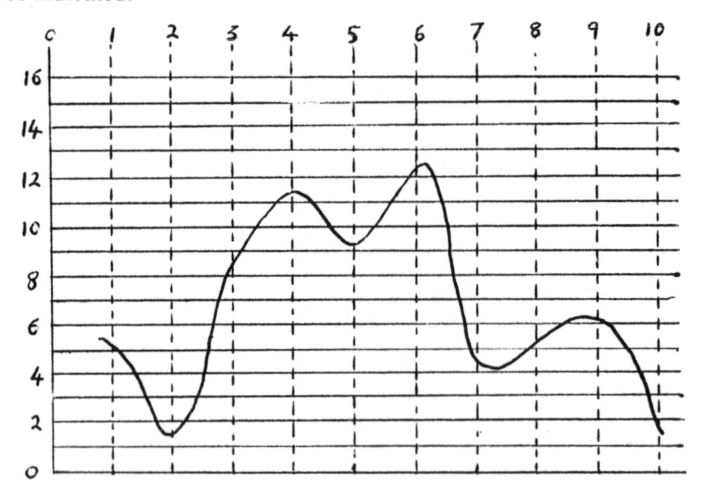

2. Reconstruct, on graph paper, the waveform which has been encoded using 16 equally spaced amplitude levels at ten equal intervals, using p.c.m. A parity check digit has been appended to each codename to make ten protected code names of equal weight, the final signal being

   10111 11000 01111 00110 10001 11011 01111 10100 00011 01100.

After working through this chapter you should:

1. understand the terms parity digit, repetition code, weighted code and pulse code modulation (p.cm.);
2. be able to assess the protection given by a binary code, in terms of the numbers of errors detected and corrected;
3. be able to calculate or to verify the check digit of a weighted codename using progressive digiting;
4. understand the form of International Book Numbers;
5. understand Hamming's method of constructing economical error-correcting codes and be able to explain the result that the number of message digits is at most $2^c - 1 - c$.
6. know how to use p.c.m. to render a waveform into binary codenames.

Weighted codes : ISBNs

CHANNEL ENCODING

TASKSHEET 1

An error which is not caught by a simple parity check is the loss of a digit in transmission, or the insertion of an extra one. In systems involving humans there are also the common errors of misreading or mishearing digits. The most common misreadings are illustrated by these examples.

79 read as 97
443 read as 433

In the alphabet/number system the letter O and the figure zero are often confused.

For these reasons, particularly when human error is possible (e.g. in dealing with credit cards or inventory labels), a **weighted code** is used. In such a code, starting from the right, the digits are given the weights 1, 2, 3, ... , the first being the check digit which invariably has weight 1. Each digit is multiplied by its weight and the sum is found, the check digit being chosen so that the sum is divisible by some modulus. The procedure is shown in the following example using modulus 5.

codename	3	2	2	x
	x	x	x	x
weight	4	3	2	1

sum $\quad 12 + 6 + 4 + x \quad = \quad 22 + x$

x is the check digit to be found modulus 5.

x is chosen as the least positive number such that $(22 + x)$ is a multiple of the modulus 5. This multiple is 25, so $x = 3$ and the protected codename is 3223.

In a weighted code system a prime modulus is always used and the digits are all less than this modulus. This may explain some mysteries in the example above. The reason for the choice of a prime modulus is explored later, in questions 3 and 4. For codes using decimal digits the modulus 11 is normally used; for alphabet - number codes, 37 is chosen.

1. Explain the choices of 11 and 37.

For calculations in the alphabet-number code each digit takes a numerical value:

0 = 0, 1 = 1, ... , 9 = 9, A = 10, B = 11, ... , Z = 35, space = 36.

(continued)

With lengthy labels the computation of the check digit could involve some heavy arithmetic so a weighting of r is achieved by adding r times rather than multiplying by r, using a process called 'progressive digiting'. For example to encode

F 35 2

	sum	running total
F = 15	15	15
3 = 3	18	33
5 = 5	23	56
space = 36	59	115
2 = 2	61	176
check digit $x = x$	$(61 + x)$	$(237 + x)$

$$237 + x \equiv 15 + x \pmod{37}$$

$$\Rightarrow \quad x = 22 = \mathrm{M}$$

Protected codename : F35 2M

All books are now given an ISBN (International Standard Book Number), usually printed on the same page as the publisher's name and the date of publication . Items of computer software are also labelled with ISBNs. The ISBN uses decimal digits and is protected by a weighted code (mod 11). The style of an ISBN is

group identifier/space or hyphen/publisher prefix/space or hyphen/title number/ space or hyphen/check digit.

The group identifier uses from one to five digits and identifies the area of publication. The UK uses the single digit group identifiers 0 and 1. The second group consists of from one to seven digits and identifies the publisher. Title numbers have from one to six digits.

The 16-19 Mathematics book *Foundations* has ISBN

group identifier	publisher prefix	title number	check digit
0	521	38842	2

Check of validity:

ISBN	0	5	2	1	3	8	8	4	2	2
Weight	10	9	8	7	6	5	4	3	2	1

Products $0 + 45 + 16 + 7 \quad + \quad 18 + 40 + 32 + 12 + 4 \quad + \quad 2$

Total : $176 = 16 \times 11$

Note that if the method requires a check digit representing 10 then the Roman numeral X is used. X is not used in any other position.

(continued)

2. In the weighted code source {0, 1, 2, 3, 4, 5, 6} (mod 7), what check digit should be appended to the message group 40316?

3. Show that in the code of question 2 the check will always succeed (i.e. error will be detected) if

   (a) any two adjacent digits have been interchanged, or

   (b) any one digit has been wrongly copied.

4. Show the importance of using a prime modulus, by examining the case of the message group

   2704463,

   using a weighted code (mod 8) when the sequence 446 is wrongly transcribed as 466.

5. Show how the method of progressive digiting would result in the correct weighting if applied to the codename

   $a \quad b \quad c \quad x$

   where a, b, c and x are digits in the range 0 - 4.

6. Explain the step in the progressive digiting example:

   $$237 + x \equiv 15 + x \pmod{37}.$$

7. A very large UK publishing firm wants to have more than 100 000 items in its list. How many digits can it allow for the publisher prefix? How many items can it then have in its list?

8. Find the check digit x for the book with ISBN

   1 869931 02 x

   What can you say about the publishers of this book?

9. The following ISBN appears in a library catalogue

   1 – 584 – 72043 – X

   Show that is is faulty and suggest what might have caused the error.

Tutorial sheet

CHANNEL ENCODING

TASKSHEET 2

1. One way of making efficient use of check digits is to form a rectangular block of the message digits and then append a check digit to each row and column. The result is a protected **rectangular code**

```
x   x   ...   x  |  x  }
x   x   ...   x  |  x  }  check digits
.   .   ...   .  |  .  }  for rows
x   x   ...   x  |  x  }

x   x   ...   x
check digits for columns
```

A protected rectangular code

(a)
```
0 1 0 0 1 1  1
1 1 0 1 0 1  0
1 0 1 0 1 1
0 0 0 1 1 1
0 0
```

The diagram shows a 4 x 6 message block together with four of the check digits. Calculate the remaining check digits.

(b) A 36-bit message is to be encoded using a protected rectangular code. Find the most efficient way to arrange the 36 bits and show that it is the most efficient. Generalise on your conclusion.

(c) (i) An $m \times n$ message block is protected by row and column check bits. What is the efficiency of the resulting code?

(ii) Explain why a further digit in the bottom right-hand corner to serve as a parity check on row check digits also acts as a check on column check digits. What is the efficiency of the resulting code?

2.

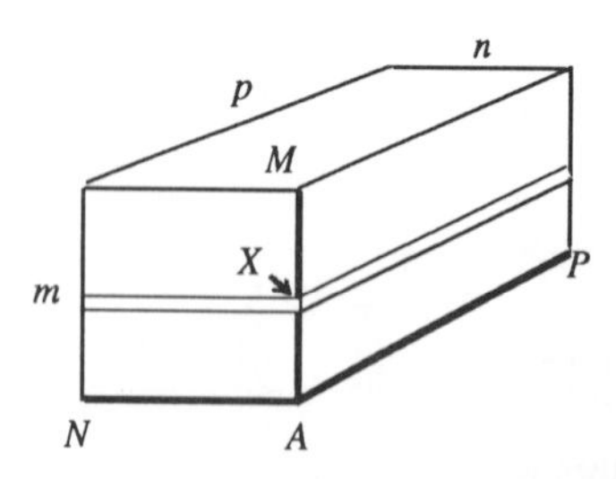

The idea of a rectangular code may be extended into three dimensions; an $m \times n \times p$ cuboidal message group is shown.

The plane array shown as a horizontal slice is protected by a **single** bit at X as described in part (ii) of Question 1(c). The complete block is protected by $m + n + p$ such bits along the edges AM, AN, AP together with a further bit at A.

(a) Explain how a further bit at A checks all three sets of check bits along these three edges.

(b) A 64-bit message can be encoded as a square or as a cube, then protected by check digits, including a corner digit in each case. Compare the efficiencies of the two methods.

5 The Mariner-9 code

5.1 Hamming distance

R.W. Hamming

R.W. Hamming, in 1950, was the first to publish work on the concept of the distance between codenames and the use of this idea in error correction. To take a simple example, among the codenames of length two:

11, 10, 01, 00,

the pair 11 and 10 differ in only one position whereas the pair 11 and 00 differ in both positions.

Following Hamming's definition you can say that the **distance** between the first pair is 1 whereas that between the second pair is 2; symbolically,

$$d(11, 10) = 1, \quad d(11, 00) = 2.$$

The 'natural' two-bit code to use to distinguish between two messages would be either (10, 01) or (11, 00), already familiar as R 2. In both of these the Hamming distance is 2.

A geometrical view of the situation reinforces the notion of distance. Taking the digits as coordinates a square is formed and you can see that the codenames for the 'better' codes represent diagonally opposite points, as far apart as possible in the square. The idea may be extended into three dimensions without much difficulty.

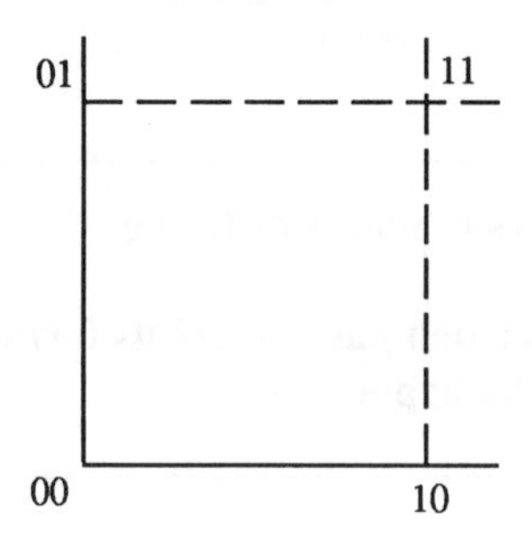

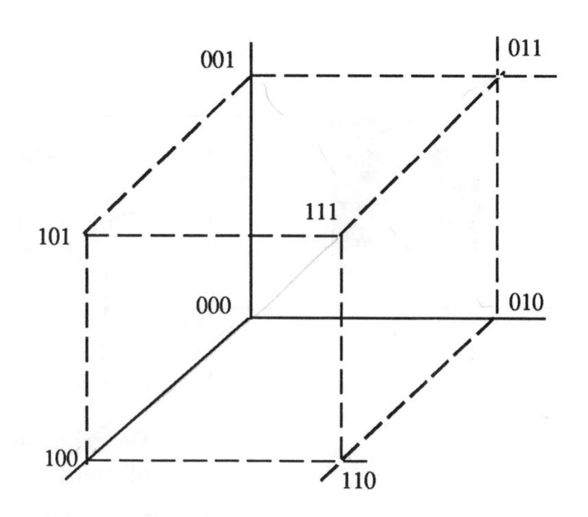

In the second diagram, points differing in three coordinates are diagonally opposite in the cube and those differing in two positions are diagonally opposite in its square faces. Thus

$$d\,(001, 110) = 3$$

indicates that the points given by 001 and 110 are as far apart as possible in the cube.

$$d\,(101, 011) = 2$$

shows that the corresponding points are as far apart as possible in a square face. Codenames with a Hamming distance of 3 are in pairs whereas those with a Hamming distance of 2 are in quadruples.

From the diagram find the three points *p* such that

$$\mathbf{d\,(101,}\ p\mathbf{) = 2.}$$

These three points, together with 101, illustrate the geometrical property of **skewness**. Two lines in space are said to be skew if they are not parallel and yet do not intersect.

(a) Identify the pairs of skew lines given by this quadruple.

(b) Find all other quadruples of 3-bit codenames such that the Hamming distance between any pair is 2.

The geometrical idea can be extended into four dimensions by considering a **hypercube**. To draw a 4-dimensional object on a sheet of paper is tricky, but the usual sort of representation is as shown.

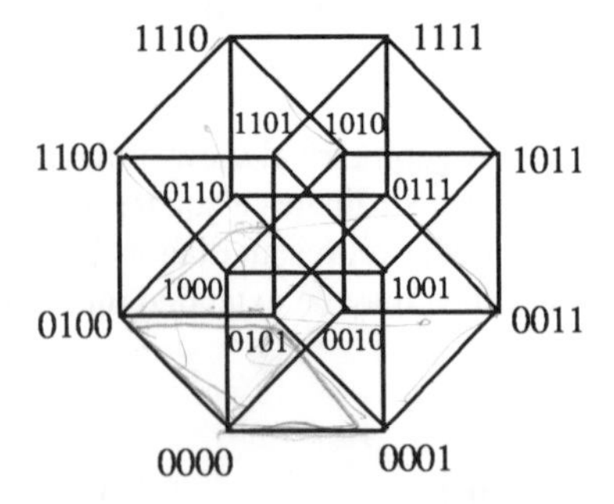

From the diagram, find pairs of points differing

(a) in all four coordinates (such pairs of points form the ends of a diagonal of the hypercube);

(b) in three coordinates;

(c) in two coordinates.

(d) Describe geometrically the line defined by the pairs of points in cases (b) and (c).

5.2 (n, k, h) codes

For any given code, if the codenames are taken in pairs, the minimum Hamming distance for the code is the least of the distances between the pairs. For example, in the 2-in-5 code

{00011 , 00101 , 01001 , 10001 , 00110 , 01010 , 10010 , 01100 , 10100 , 11000},

the distance between some pairs of codenames is 4 [e.g. d (00011, 01100) = 4)] and the distance between other pairs is 2 [e.g. d (00011, 00101) = 2]. The least distance is therefore 2.

> **(a) Explain why the distance between any pair of 2-in-5 codenames must be at least 2.**
>
> **(b) Find the minimum distance between any two codenames in Rh.**

The importance of the minimum distance lies in its relationship with the error protection given by a code. In Chapter 4 you obtained the following result for Rh:

Code	Minimum distance	Errors		
			detected	corrected
Rh	h	h even	$\frac{1}{2}h$	$\frac{1}{2}(h-2)$
		h odd	$\frac{1}{2}(h-1)$	$\frac{1}{2}(h-1)$

In fact, the following result can be proved for any channel encoding:

> **If h is the minimum distance for a code then**
>
> - **if h is odd the code can detect and correct up to $\frac{1}{2}(h-1)$ errors;**
> - **if h is even the code can detect up to $\frac{1}{2}h$ errors and correct up to $\frac{1}{2}(h-2)$ errors.**

A proof of the general result is not difficult and is given in Tasksheet 1E.

TASKSHEET 1E - *Minimum distance*

> **What error protection is given by a code with minimum distance 5?**

The minimum Hamming distance is often represented by the letter h and an (n, k) code with minimum distance h is classified as an (n, k, h) code.

> **An (n, k, h) code is any binary code such that**
> - **codenames have n digits, k of which are message digits**
> - **the minimum Hamming distance between codenames is h.**

For example, {001, 110} is a (3, 1, 3) code.

> **Classify the code consisting of all 4-bit codenames.**

Hamming developed a special technique for constructing $(2k, k, k)$ codes. His method may be illustrated in the case $k = 3$.

Construction of a (6, 3, 3) code:

Let x_1, x_2 and x_3 be message bits. The codename

$$x_1 x_2 x_3 x_4 x_5 x_6$$

is obtained using check digits x_4, x_5 and x_6 chosen such that the sums

$$\begin{array}{llllll} & x_2 & + x_3 & + x_4 & & \\ x_1 & & + x_3 & & + x_5 & \\ x_1 & + x_2 & & & & + x_6 \end{array}$$

are all even.

Note that the first check digit x_4 checks all message digits except the first; the second check digit x_5 checks all message digits except the second and similarly for the third check digit. Thus if the message group is 101 with $x_1 = 1$, $x_2 = 0$ and $x_3 = 1$ then the check group is $x_4 x_5 x_6$ where

$$\begin{array}{llllll} & 0 & + 1 & + x_4, & & \\ 1 & & + 1 & & + x_5, & \\ 1 & + 0 & & & & + x_6 \end{array}$$

are all even; from which you can see that $x_4 = 1$, $x_5 = 0$, $x_6 = 1$ and the complete codename is 101101.

> **Use this method to find the other 7 codenames in the Hamming (6, 3, 3) code.**

Exercise 1

1. (a) Explain why the code {000, 011, 101, 110} is a (3, 2, 2) code.

   (b) Find another (3, 2, 2) code.

2. (a) Using Hamming's procedure construct

   (i) the (4, 2, 2) code, (ii) the (8, 4, 4) code.

   (b) What error protection is given by these two codes?

3. The Hamming (10, 5, 5) code, with I.T.A. to decode the message group, was used by the Marconi Company under the name AUTOSPEC (automatic single path error correction).

   (a) In AUTOSPEC what are the check groups for the message groups

   (i) 01101, (ii) 00110 ?

   (b) Suppose you check an AUTOSPEC codename by forming a 5-letter error word as follows. If the first check digit appears correct, write Y in the first position. If it appears incorrect, write N in the first position. Similarly, writeY or N in the remaining positions according to whether or not the corresponding check digit appears correct. Thus if you receive the signal

   11010 00000

   the error word will be YYNYN. If you receive the signal

   11101 10101

   the error word will be YNYYY.

   (i) What type of error word would be produced if one bit of the message group had been transmitted incorrectly?

   (ii) What type of error word would be produced if one bit of the check group had been corrupted?

   (c) (i) Write out the error words for the signals

   11010 00100 and 10010 00101

   (ii) Explain what corruption in the signal gave rise to these error words.

   (iii) On the basis of your answer to part (ii), describe how a machine decoder might process AUTOSPEC signals.

With a partner, practise encoding and decoding messages using a Huffman - AUTOSPEC system, i.e. encode using Huffman, break into 5-bit strings and then apply AUTOSPEC protection.

5.3 Metric spaces

In Section 5.1 the geometrical idea of distance was used to reinforce the concept of the Hamming distance between codenames.

Geometrical vectors can be added together or multiplied by scalars.

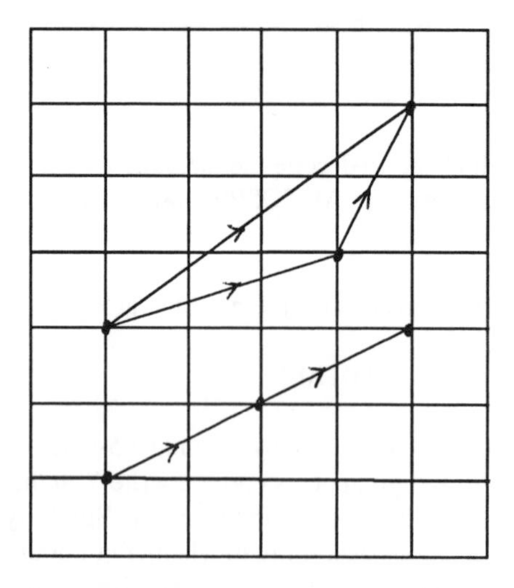

$$\begin{bmatrix} 3 \\ 1 \end{bmatrix} + \begin{bmatrix} 1 \\ 2 \end{bmatrix} = \begin{bmatrix} 4 \\ 3 \end{bmatrix}$$

$$2\begin{bmatrix} 2 \\ 1 \end{bmatrix} = \begin{bmatrix} 4 \\ 2 \end{bmatrix}$$

The set of vectors of any given dimension, together with the operations of addition and multiplication by scalars, form an algebraic structure called a **vector space**. Since the usual geometrical distance function (or metric) can also be applied to these vectors, this particular vector space is called a **metric space**.

Codenames are vectors and the set of codenames of a given length can also be considered to be a metric space. The Hamming distance function forms the metric and the addition and scalar multiplication of codenames is covered on Tasksheet 2.

TASKSHEET 2 – *Linear combinations of codenames*

The complete code C_n with bitwise addition (mod 2) and the Hamming distance function d forms a metric space.

A **spanning set** is a set of elements of a metric space whose linear combinations generate every element of the space. This idea is needed for Section 5.4 on the Mariner-9 code.

Example 1

Show that the codenames 01 and 11 span C_2.

Solution

Let $\mathbf{s}$ = 01 and $\mathbf{t}$ = 11. Then

$$0\mathbf{s} + 0\mathbf{t} = 00$$
$$1\mathbf{s} + 0\mathbf{t} = 01$$
$$0\mathbf{s} + 1\mathbf{t} = 11$$
$$1\mathbf{s} + 1\mathbf{t} = 10$$

So {01, 11} is a spanning set for C_2.

Addition of binary codenames involves adding corresponding bits modulo 2, for example

1010 + 0110 = 1100

A spanning set for a space is any set of codenames whose linear combinations generate every element of the space.

Exercise 2

1. (a) Show that all codenames of length 3 are generated as linear combinations of

   010, 110 and 111

   (b) Find another set of three codenames spanning C_3.

   (c) Show that a basic set of only two members cannot generate C_3.

   (d) Suppose that $\mathbf{r}$, $\mathbf{s}$ and $\mathbf{t}$ are three non-zero codenames of C_3 which do **not** span C_3. How are $\mathbf{r}$, $\mathbf{s}$ and $\mathbf{t}$ related?

2. Express the codename 0110 as a linear combination of the codenames:

   (a) 1100, 0101, 1011, 1111

   (b) 0100, 0101, 1011, 1111

5.4 Space science

Space exploration, such as the Mariner and Viking missions to Mars and the Giotto probe to Halley's Comet, produces huge amounts of data for transmission to scientists on Earth. Just being able to take observations from outside the Earth's atmosphere enables astronomers to gain fuller and more accurate information about the universe. For example, the European Space Agency's High Precision Parallax Collecting Satellite (Hipparcos) will measure changes in the positions of the stars with unprecedented precision, enabling astronomers to calculate stellar distances and thus to learn more about the physics of the universe and its origins.

The Hipparcos mission illustrates many of the ideas of this unit. Hipparcos will analyse the positions and motions of some 120 000 stars. Light received by the satellite is focussed onto a modulating grid, enabling the light intensity to be sampled over 1000 times each second. Digital information will be transmitted back to Earth at the rate of 24 000 bits per second over a $2\frac{1}{2}$ year period. The analysis of this data is itself an immense undertaking. Preparatory work started in 1982 and by about 1995 the results of the mission will be available to the scientific community.

The original Hipparchus, a Greek astronomer (190-120BC), used parallax to determine the distance of the Moon from Earth.

As an example of the amount of protection needed when transmitting data back to Earth, the channel encoding of the Mariner-9 signals is considered in the next tasksheet.

TASKSHEET 3 – *Mariner-9*

5.5 Spreading the message

In the last section you considered some of the issues concerned with encoding and transmitting data back to Earth. Deciding which data should be transmitted into space poses other problems.

Suppose you had to devise a message to be put into a space-probe and sent off on a galactic voyage in the hope that the vessel might be intercepted and the message found by some far-off intelligent creature. What whould that message look like? The problem resembles that of deciding what to write on a bit of paper you put into a bottle before throwing it into the sea or fasten to a helium balloon before releasing it into the air. For instance, you would want to include your address and possibly tell the recipient something about yourself. A message actually sent into space is shown here.

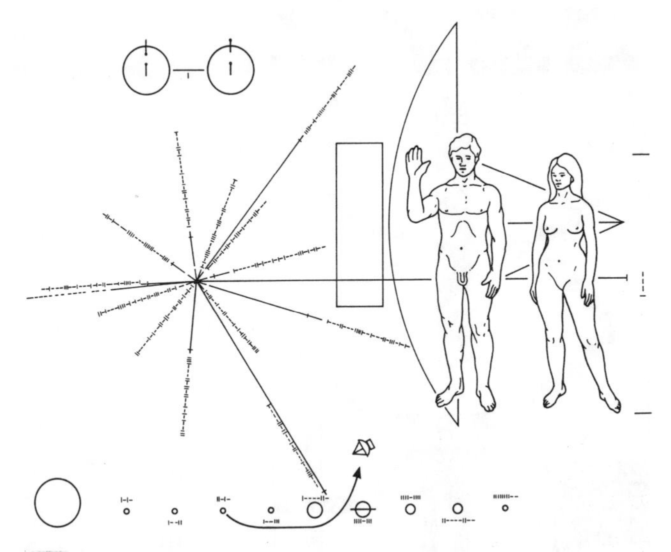

These diagrams, inscribed on a plaque, were sent in Pioneers 10 and 11. Top left is a representation of a rotating hydrogen atom. At the bottom are the Sun and planets, Earth being indicated as the source of the probe. Behind the man and woman is a simplified sketch of the probe. The remaining object shows the directions of the principal pulsars from Earth, together with their periods.

The pulsars referred to in the caption are sources within our galaxy of regular radio signals, discovered in 1967. Their emissions are so apparently meaningful that astrophysicists humorously designated pulsars LGM, standing for 'little green men'. Since the time of their discovery various listening projects have been designed, the first under way being Project CETI (Communication for Extra-Terrestrial Intelligence) initiated by the Soviet Union following a Soviet-American conference organized by Carl Sagan in 1973. To date, no message has been decoded from the myriad radio signals reaching Earth from our galaxy.

The first signal to be transmitted from Earth to attempt to establish contact with an extra-terrestrial intelligence was sent by the Arecibo radio telescope in Puerto Rico, in 1974. This signal depended on the fact that the number 1679 has only two prime factors and so a signal of 1679 binary digits could be rendered into a rectangular display in only one way, as a 23 x 73 array. If O's are replaced by white squares and 1's by black squares, the picture overleaf is obtained.

Message transmitted by the Arecibo radio telescope

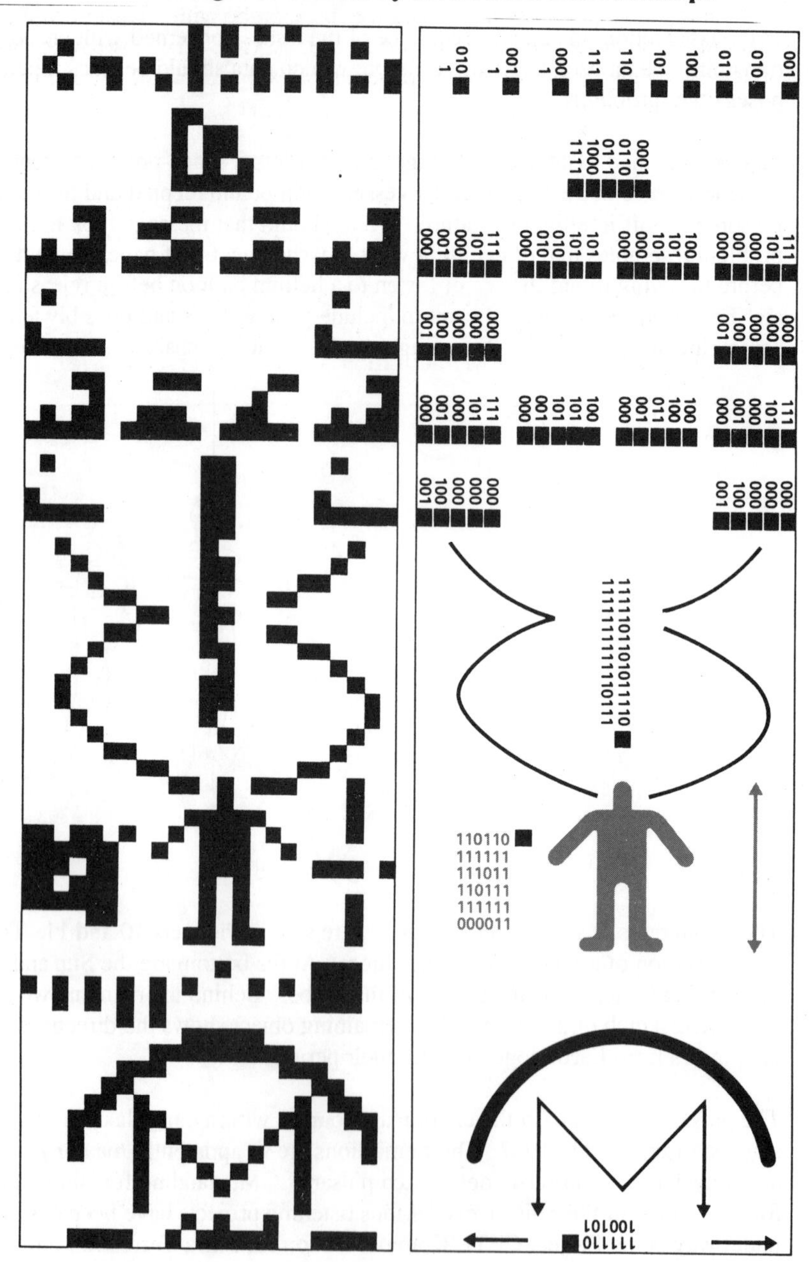

The picture on the left is the image obtained by using white and black squares. On the right is a form which is easier to understand.

The picture is intended to be interpreted as follows. At the bottom is the Arecibo radio telescope. Above this are shown the sun and planets, the third planet Earth being offset. Then comes a human form with a representation of the double helix of DNA and finally there are chemical formulas for the compounds composing the DNA molecule.

In general, how can messages be recognised? All messages seem to have two aspects. Consider for example a town crier, who rings a bell before intoning his notices, a book containing long strings of letters and the space probe with its inscribed plaque. All exhibit the bipartite and aperiodic nature of messages. The first part is some device to catch the attention; the second part, which follows or is found within the first, reveals actual content in a non-repeating string of characters.

(a) How can you distinguish between a randomly generated string of characters and one intended to convey a message?

(b) Messages sometimes appear to be periodic, for example SOS or 'Mayday' distress calls, which are repeated many times. Does this contradict the definition of messages given above?

After working through this chapter you should:

1. understand the terms Hamming distance, linear combination, spanning set and the notation (n, k, h) to describe a binary code;
2. be able to interpret Hamming distances geometrically;
3. be aware of the relationship between minimum Hamming distance and error protection;
4. know how to construct $(2k, k, k)$ codes, including the AUTOSPEC (10, 5, 5) code;
5. know how to construct the codes H_n, including the Mariner-9 code H_5.

Minimum distance

The following theorem is especially important for the theory of error correction:

If the minimum distance h for a code is odd then the code can detect and correct up to $\frac{1}{2}(h-1)$ errors. If h is even then the code can detect up to $\frac{1}{2}h$ errors and correct up to $\frac{1}{2}(h-2)$ errors.

One explanation uses the idea that d has a property typical of distance measures in that it satisfies the **triangle equality**: if x, y and z are any three codenames then

$$d(x, z) \leq d(x, y) + d(y, z).$$

1. Explain why the triangle equality is satisfied by d.

Suppose that the codename a is transmitted and that it is corrupted by n errors. The result is a binary group x with $d(x, a) = n$. The idea of error correction is to replace x by the codename nearest to x.

Now suppose that b is any other codename, i.e. $b \neq a$. From the definition of h as the minimum distance between codenames,

$$d(a, b) \geq h \qquad ①$$

From the triangle inequality,

$$d(a, b) \leq d(a, x) + d(x, b). \qquad ②$$

2. By combining inequalities ① and ②, show that

$$d(x, b) \geq h - n$$

for **any** codename b different from a.

If $n < h - n$ then a is the codename nearest x and the correct codename will be found by replacing x by the nearest codename.

3. Complete the proof in the case when h is odd.

4. Complete the proof in the case when h is even.

Linear combinations of codenames

THE MARINER-9 CODE

TASKSHEET 2

Addition of binary codenames of the same length means addition of corresponding bits of the codenames (modulo 2). The complete addition table modulo 2 is

$$0+0=0, \quad 0+1=1+0=1, \quad 1+1=0$$

1. Show that $111 + 101 = 010$.

2. Interpret **subtraction** of codenames and show that

   $$111 - 101 = 010.$$

   Make a general statement about the relation between addition and subtraction modulo 2.

3. Evaluate (i) $1011 + 1011$ (ii) $1011 + 1011 + 1011$.

   Make a general statement about multiples of a binary codename.

4. Let $\mathbf{r} = 1010$, $\mathbf{s} = 0011$, $\mathbf{t} = 1110$. Calculate

   (i) $\mathbf{s}+\mathbf{t}$ (ii) $\mathbf{r}+\mathbf{t}$ (iii) $\mathbf{r}+\mathbf{s}+\mathbf{t}$.

A **linear combination** of **r**, **s** and **t** is any codename of the form

$$a\mathbf{r} + b\mathbf{s} + c\mathbf{t},$$

where a, b and c are integers.

5. Explain why the only integer values which need be considered for a, b and c are 0 and 1.

6. Show that there are 8 possible linear combinations of **r**, **s** and **t**.

7. Construct the complete set of linear combinations of

   10101, 00110, 00011 and 10000.

   Comment on your answer.

Mariner-9

THE MARINER-9 CODE

TASKSHEET 3

The Mariner space missions to Mars, especially the Mariner-9 mission in 1971-1972, enabled scientists to construct the first detailed maps of the Martian surface. This prepared the way for the later Viking mission to land two automatic stations which were then able to analyse the landscape with cameras and also to analyse ground samples.

Part of the Martian surface.

Transmissions from the Mariner-9 spacecraft of encoded pictures of Mars were basically of signals in 9-bit strings. These strings were repackaged into 6-bit groups (i.e. two 9-bit codenames became three 6-bit groups) and each 6-bit group was sent with 26 bits of protection as a 32-bit string.

The Mariner-9 code is a member of a sequence of codes given the names H_0, H_1, H_2, H_3, H_4, H_5, ... All codes in the sequence $\{H_n\}$ are formed in the same way, by finding all linear combinations of a basic set of codenames.

The basic sets

A basic set for H_n consists of

$$\underbrace{0 \ldots 0}_{2^{n-1}\text{ 0's}} \quad \underbrace{1 \ldots 1}_{2^{n-1}\text{ 1's}}$$

together with two-fold repetitions of the codenames in a basic set for H_{n-1}.

H_0 has basic set 1

$\Rightarrow H_1$ has basic set 01, 11

$\Rightarrow H_2$ has basic set 0011, 0101, 1111

etc.

(continued)

Construction of H_2

The codenames of H_5 are very long and so H_2 will be used to illustrate the method of construction. H_2 consists of all linear combinations of

$$x_1 = 0011, \quad x_2 = 0101, \quad x_3 = 1111$$

$$\begin{aligned}
\mathbf{0} &= 0000 \\
x_1 &= 0011 \\
x_2 &= 0101 \\
x_1 + x_2 &= 0110 \\
x_3 &= 1111 \\
x_1 + x_3 &= 1100 \\
x_2 + x_3 &= 1010 \\
x_1 + x_2 + x_3 &= 1001
\end{aligned}$$

H_2 is the even-weight code of length 4 already described in Chapter 4. You will have found that it gives only single-error detection.

1. Construct the code H_1.

2. H_3 has four basic codenames. Since each of these may be either excluded or included in any linear combination and the two choices may be made independently there are $2 \times 2 \times 2 \times 2 = 16$ codenames in H_3. List these codenames.

3. For the code H_4, how many members has

   (a) the basic set (b) the code?

4. What is the minimum Hamming distance between members of

   (a) H_1 (b) H_2 (c) H_3?

5. Find the error protection given by the codes H_n for $n = 1, 2, 3, 4$ and 5.

6. (a) Show that the code C_3 is **embedded** in H_2 **systematically**, in that the 1st, 2nd and 4th digits of the groups in H_2 give the codenames of C_3.

   (b) How would you describe H_2 using the (n, k, h) notation?

7. (a) How is C_4 embedded in H_3?

   (b) How would you expect the message bits to be embedded in the Mariner-9 code?

   (c) Classify H_n using the (n, k, h) notation.

Tutorial sheet

1. Consider the set of four 4-bit codenames

$$S = \{1011, 1101, 0110, 0011\}.$$

(a) Express one of these codenames as a linear combination of the others.

(b) Find three different subsets of S, each with three members, which span the same set of codenames as S.

(c) Using the subset $T = \{1011, 1101, 0011\}$, how many codenames can be generated by linear combinations? Express 1000 as such a linear combination.

2. Frank Drake used a number (1679) uniquely expressible as a product of two primes (23 x 73) to devise a rectangular block of bits signalling a message in the form of a diagram. Imitate this idea to signal a sample message of your own choice, e.g. the shape of a capital letter.

3. [A knowledge of the multiplication law for probabilities is needed for this question.]

The second extension of a source S is written S^2 and is defined as ordered pairs of members of S. For example, suppose S is:

Member	a	b
Probability	$\frac{3}{4}$	$\frac{1}{4}$

then S^2 is :

Member	aa	ab	ba	bb
Probability	$\frac{9}{16}$	$\frac{3}{16}$	$\frac{3}{16}$	$\frac{1}{16}$

(a) Find T^2 if T is the source:

Member	x	y	z
Probability	$\frac{1}{2}$	$\frac{1}{4}$	$\frac{1}{4}$

(b) Find $H(S)$, $H(S^2)$, $H(T)$, $H(T^2)$ and comment on your results.

SOLUTIONS

1.3 Binary encoding of information

What questions might you ask? How many questions will you need to ask?

You could simply guess objects at random.

For example,

'Is it the cone?'

'Is it the sphere?' etc.

One, two or three questions of this form would be needed, depending on how lucky you were.

When finding an object using yes/no questions, you might be lucky with a pure guess. However, if you are unlucky then you have eliminated only one possibility and you are very little further forward. A more refined method is to attempt to halve the remaining probabilities at each stage by devising suitable questions. In general this proves to be the most efficient strategy.

Exercise 1

1. (a) The first question halves the list, leaving 4.
   The second question halves the remaining list, leaving 2.
   The third question identifies the card.

   (b) One possible set of three questions:

   1. Is the shape shaded?
   2. Is it curved?
   3. Has it more than one line of symmetry?

(c)

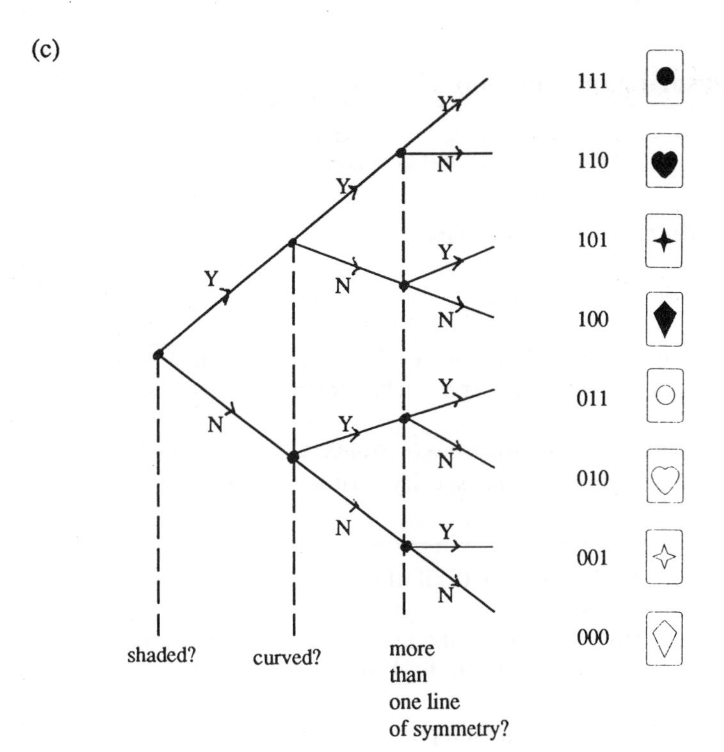

2. (a) $2^4 = 16$

   (b) First digit : male 1, female 0;
   Second digit : mature 1, immature 0;
   Third digit : full plumage 1, otherwise 0;
   Fourth digit : on ground 1, perching 0.

   The bird described would be labelled 0110.

3. (a) Each quadrant could either contain a dot or be left empty. Then we have $2^4 = 16$ possible signs. One way of interpreting as a binary system is shown in these examples.

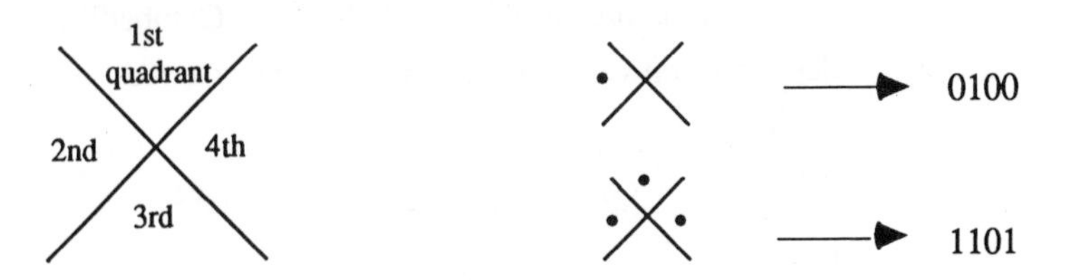

   (b) The most likely explanation is that it is a direction sign left in a 'trail-blazing' exercise by army cadets or scouts.

4. $2^8 = 256$. Instructions are in fact often given in 8-bit 'bytes'.

Possible project

For this project you would need to consider alternative strategies for 'guessing'. Some knowledge of probability theory would be useful.

1.4 Measuring information

If $n(S) = 2^k$, what is H(S)? Justify your answer.

Consider all sequences of k bits:

0 or 1   0 or 1   ...   0 or 1.

There are two possibilities for each bit and so there are $\overbrace{2 \times 2 \times \ldots \times 2}^{k \text{ factors}} = 2^k$ possible sequences. Each member of S can be specified using k bits and so H(S) = k.

This result holds even when $k = 0$. If $n(S) = 2^0$, then the set has only one number and no information is needed to specify a particular member i.e. H(S) = 0.

(a) Find H(S) if $n(S) = 12$.

(b) What is the average number of questions needed to find somebody's month of birth?

(c) Suggest a possible set of questions.

(a) $\text{Log}_2 12 \approx 3.585$.

(b) Four months would be found using 3 questions. The remaining eight months would need 4 questions. The average number is

$$\frac{(4 \times 3) + (8 \times 4)}{12} = 3\frac{2}{3}.$$

(c) The questions might start with

'Is it before July?'

followed by

'Is it before April?' or 'Is it before October?'

depending on whether the answer to the first question is 'yes' or 'no'.

Exercise 2

1. (a) 5.

(b) A binary decision tree gives 20 tips at the end of 5 branches and 6 tips at the end of 4 branches i.e.

5 questions are needed in 20 cases,
4 questions are needed in 6 cases.

So the average number of questions is

$$\frac{(20 \times 5) + (6 \times 4)}{26} = 4.77.$$

(c) $\log_2 26 = 4.70$.

2. The least positive integer power of 2 which is greater than 100 is $2^7 = 128$. So
 6 questions will be needed in (128–100) = 28 cases,
 7 questions will be needed in (100–28) = 72 cases.
 The average number of questions is

$$\frac{(72 \times 7) + (28 \times 6)}{100} = 6.72.$$

3. (a) The blank tile would be drawn, on average, every 27th time. So the average length of a 'word' would be 26 letters.

 (b) $\dfrac{(22 \times 5) + (5 \times 4)}{27} = 4.81$ (questions).

 (c) $4.81 \times 5 = 24.1$ (questions).

 (d) $5 \log_2 27 = 23.8$ (bits). [Possibly $5 \log_2 26$ if you omit the space tile]

4. If you assume that all the dots may take any degree of brightness with equal probability, the information content would be

$$600 \times 800 \times \log_2 10 = 1.59 \times 10^6 \text{ (bits)}.$$

 Assuming the result in question 3(c), a thousand words would have information content of about 2.38×10^4; very much less than that of the TV picture.

 In fact, neither of these measures is accurate because equal probabilities have been assumed without justification. However, even in a more refined model you should still expect a TV picture to be 'worth' many thousands of words.

Possible project

Your answer should include a drawing of each letter of the alphabet as an array of tubes, together with a binary representation according to whether tubes are on or off in its display. Example: if tube no. 1 is off, tube no. 2 is on, tube no. 3 is on ... the code-name for the letter would start 011... .

Some measure of the efficiency of the system would involve comparing the theoretical minimum number of tubes for a 26 letter alphabet with the actual number you use. A progression from economy to legibility to aesthetic appeal might be considered.

2. *Source encoding*

2.1 Unequal frequencies

Check that both methods give the value – 2.837.

Method 1 $\log_2 14 - \log_2 100 = 3.807 - 6.644$ (from Table 1)

$= -2.837$

Method 2 $\dfrac{\log_{10} 0.14}{\log_{10} 2} = \dfrac{-0.8539}{0.3010} = -2.837$

Check the above calculation using either of the recommended methods for calculating logarithms.

The terms are respectively 0.216, 0.332, 0.411, 0.464, 0.464, 0.521.

Their sum is 2.408, to 3 decimal places.

Explain the other terms in the expression.

In a 'typical sample' of 8 letters,

2 would be B's , each needing 2 questions,
1 would be a C, needing 3 questions,
1 would be a D, again needing 3 questions.

Exercise 1

1. A 'typical' sequence of 16 letters would contain

   8 A's, 4 B's, 2 C's, a D and an E.

   The information (total number of questions) for the sequence of 16 letters would be

   $(8 \times 1) + (4 \times 2) + (2 \times 3) + (1 \times 4) + (1 \times 4) = 30$

   So the average number of questions per letter is $\frac{30}{16} = 1\frac{7}{8}$

   The same answer can be obtained by applying the formula for H(*S*).

2. (a) $-(0.1 \log_2 0.1 + 0.2 \log_2 0.2 + 0.3 \log_2 0.3 + 0.4 \log_2 0.4)$

$= 0.332 + 0.464 + 0.521 + 0.529$

$= 1.85$ to 2 decimal places.

(b) 1.93 to 2 decimal places

(c) 1.99 to 2 decimal places.

The distribution giving greatest entropy is (0.25, 0.25, 0.25, 0.25), corresponding to a source in which the members have equal information content. This corresponds with the situation in thermodynamics when the positions and energies of gas molecules are randomly distributed, i.e. the system is completely disordered.

2.2 Block codes and instantaneous codes

Do you think the suggested code is well chosen?

It is clear that the given code is not the best because although A is more frequent than B its codename is longer.

Exercise 2

1. A – 11, B – 10, C – 011, D – 010, T – 001, E – 000

(a) (i) BAT (ii) BED

(b) (i) 0111110 (ii) 01000010001

2. (a)

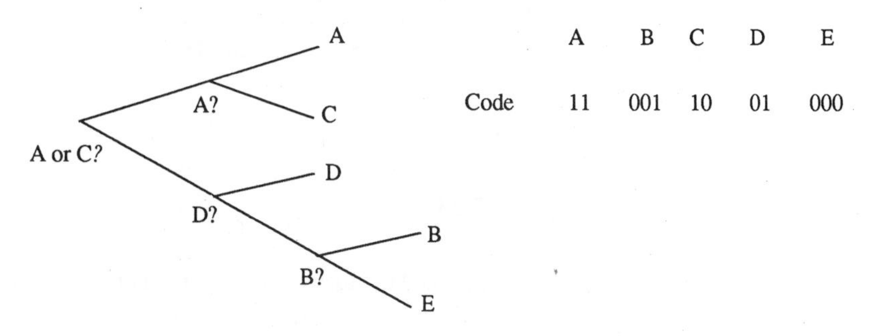

	A	B	C	D	E
Code	11	001	10	01	000

(b) $H(S) = \frac{1}{4}\log_2 4 + \frac{1}{8}\log_2 8 + \frac{1}{4}\log_2 4 + \frac{1}{4}\log_2 4 + \frac{1}{8}\log_2 8$

$$= \frac{1}{2} + \frac{3}{8} + \frac{1}{2} + \frac{1}{2} + \frac{3}{8} = 2\frac{1}{4}$$

(c) $\bar{L} = (\frac{1}{4} \times 1) + (\frac{1}{8} \times 4) + (\frac{1}{4} \times 2) + (\frac{1}{4} \times 3) + (\frac{1}{8} \times 4) = 2\frac{1}{2}$

(d) $\bar{L} \geq H(S)$ [There would be equality if the code in (a) were used.]

3. (a) A code is instantaneous if no codename is the start of another. For example, a code containing codenames 1 and 10 cannot be instantaneous. If a signal started with 10 the decoder would either automatically register A or cease to function because of the ambiguity; in either case C would never be registered.

(b) If a tree code were **not** instantaneous then one code name α would be the start of another codename β, for example $\alpha = 1$ and $\beta = 10$.

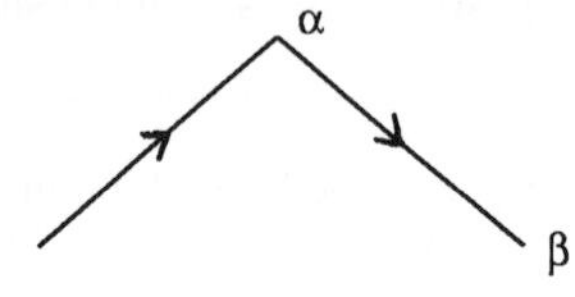

Then the sequence of branches to reach β would start with the sequence to reach α. But α labels a tip of the tree, so the sequence for β could not be completed.

4. (a) $H(S) = 2.16$

(b)

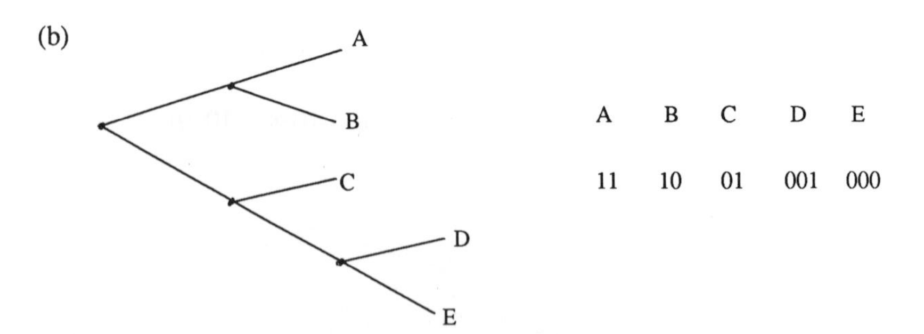

A	B	C	D	E
11	10	01	001	000

(c) $\bar{L} = 2.23$

(d) $\bar{L} \geq H(S)$. This turns out to be a general rule. The proof is too difficult for this introductory course.

2.4 A note on the psychology of alphabets

Find another method of converting $(D7)_{16}$ to binary form. Find a way of converting $(D7)_{16}$ to octal form.

A variety of answers is possible. 13 and 7 may be written separately in 4-digit binary form:

$$13 = 1101_2, \quad 7 = 0111_2$$

and then the digits expressed as a single string : 11010111.

One method of converting to octal form is as follows:

$$\begin{aligned}(13 \times 16 + 7)_{10} &= (15 \times 20 + 7)_8 \\ &= (320 + 7)_8 \\ &= 327_8\end{aligned}$$

See the multiplication table in question 4 of exercise 3

Exercise 3

1. Each octal digit is formed from a group of three binary digits.

For example:

binary number	101	010	011
octal number	5	2	3

Each hexadecimal digit represents a group of four binary digits.

decimal number	37
binary equivalent	100101
octal form	4 5
hexadecimal form	2 5

2. The decimal number 25 = (3 x 8) + 1 and would be written as 31 in octal form.

3. $(E8)_{16} = 11101000_2$

4. Multiplication table for octals

	1	2	3	4	5	6	7
1	1	2	3	4	5	6	7
2	2	4	6	10	12	14	16
3	3	6	11	14	17	22	25
4	4	10	14	20	24	30	34
5	5	12	17	24	31	36	43
6	6	14	22	30	36	44	52
7	7	16	25	34	43	52	61

Tom Lehrer, in his song '*New math*', comments that it's easy if you have no thumbs!

3 *Redundancy*

3.2 Measuring redundancy

Why? How would you obtain such a source?

You will recall that the average information per member (entropy) of a source with n members is greatest when the members occur with equal frequency. It is then $\log_2 n$. If you used the 27 symbol source by, say, putting 27 Scrabble tiles in a bag, drawing one at random, noting it and then replacing it etc., then you would have theoretically achieved equal frequency and entropy of $\log_2 27 \approx 4.755$.

Calculate $H_1(S)$

$$H_1(S) = \frac{-1}{\log_{10} 2}(0.1859 \log_{10} 0.1859 + 0.1031 \log_{10} 0.1031 + \ldots)$$

$$= 4.08 \text{ (to 2 decimal places).}$$

Exercise 1

1. (a) You saw in Chapter 2 that the entropy of this source is 1.75.

$$\text{Redundancy} = \frac{\log_2 4 - 1.75}{\log_2 4} = \frac{2-1.75}{2} = 12.5\%$$

(b) Entropy is $-(0.20 \log_2 0.20 + \ldots) = 1.95$ (to 2 decimal places).

$$\text{Redundancy} = \frac{\log_2 5 - 1.95}{\log_2 5} \approx 16\%$$

2. $$\text{Redundancy} = \frac{4.755 - 4.08}{4.755} \approx 14\%$$

3. $$\text{Efficiency} = \frac{H_1(S)}{\bar{L}} = \frac{4.08}{4.12} \approx 99\%$$

Possible project (redundancy in English)

Remember particularly that if no question is needed to 'guess' a letter then that letter conveys no information to you. This can happen in various ways, using your understanding of the sense or the rules of spelling. To take an obvious case, there is no need for a question after letter Q.

You should expect an answer for $H_\infty(S)$ very much lower than $H_1(S)$.

3.3 Digrams

> **(a) With 27 symbols, how many possible digrams are there?**
>
> **(b) Why do you not need to consider all of them?**

(a) From 27 symbols, 27^2 digrams can be formed.

(b) Many of these never occur in practice, e.g. QN and □□.

Possible project (a second approximation to English)

As it stands, this would not rank as a full-blown investigation (though the searching involved might take a long time). You could combine this project with the one suggested in Section 3.2.

4 Channel encoding

4.1 Error p. otection

Express this last condition in your own words

The average length of a codename cannot be less than the average information per symbol. This condition applies as long as the length and information are measured in the same units, usually bits.

4.2 Error detection

(a) Give the ITA codenames with a check digit giving even parity for (i) B (ii) ? (iii) 7.

(b) Decode the protected codenames.

(i) 100010 (ii) 001001 101000 (iii) 111010.

(c) What action would you take if you received the signal 111110?

(d) Under what cirumstances would this form of error protection fail?

(a) (i) 011000 (ii) 001001 011000 (changing mode) (iii) 000110.

(b) (i) C (ii) @ (iii) carriage return.

(c) If possible, ask for the message to be repeated (error registered).

(d) It fails when there is an even number of errors (2, 4 or 6). In cases when this form of protection is used, multiple errors are extremely rare. Typically, an error rate of 1 in 10^4 might be expected.

4.3 Efficiency and redundancy

What are the efficiency and redundancy of the repetition code R2?

The codenames are of length 2 with 1 message digit. The efficiency and redundancy are both $\frac{1}{2}$.

(a) Explain why the protected ITA code is a (6, 5) code.

(b) Classify the repetition code R3 in (n, k) form.

(a). A protected ITA codename has 6 digits, 5 of which carry the message.

(b) R3 is a (3, 1) code.

Exercise 1

1. (a) $\frac{k}{n}$ (b) $(s, 1)$.

2. The message is SEND SOCKS, with M instead of N and an incorrect check digit for C.

3. (a) 0000, 0001, 0010, 0100, 1000, 0011, 0101, 1001, 0110, 1010, 1100, 0111, 1011, 1101, 1110, 1111.

   The efficiency is 100%, since there is no redundancy in the form of check digits.

   (b) The complete (n, n) code has 2^n elements.

4. (a) 00011, 00101, 01001, 10001, 00110, 01010, 10010, 01100, 10100, 11000.

   There are $\binom{5}{2} = 10$ codenames, one for each decimal digit.

   Since the number of message bits is not defined the definition of efficiency given does not apply. A reasonable alternative would be to use the information content of a source with 10 equally likely members, $\log_2 10$. In this case, you would obtain $\frac{\log_2 10}{5} \approx 66\%$.

   (b) The van Duuren code has $\binom{7}{3} = 35$ codenames and may be used in much the same way as the ITA. Using the same interpretation of efficiency as before, you would obtain $\frac{\log_2 35}{7} \approx 73\%$.

4.4 Error correction

Exercise 2

1.

n	number of errors detected	corrected
2	1	0
3	1	1
4	2	1
5	2	2
6	3	2
7	3	3

Using Rn , the number of errors detected is

$\frac{1}{2}n$ if n is even,
$\frac{1}{2}(n-1)$ if n is odd.

The number corrected is

$\frac{1}{2}(n-2)$ if n is even,
$\frac{1}{2}(n-1)$ if n is odd.

2. (a) The even-weight code detects single errors only and cannot correct them.

   (b) Even-weight codenames of length 5 :

   00000 , 00011 , 00101 , 01001 , 10001 , 00110 , 01010 , 10010 ,
   01100 , 10100 , 11000 , 01111 , 10111 , 11011 , 11101 , 11110.

   This also detects single errors, only.

   (c) (i) The even-weight code of length n has 2^{n-1} members.

   (ii) All even-weight codes give single-error detection only.

4.5 Hamming's error-correcting system

(a) Show that the codename 1111111 is correct.

(b) Correct and decode the faulty codename 1001111.

(a) Check 1 $1+1+1+1$ (even)
Check 2 $1+1+1+1$ (even)
Check 3 $1+1+1+1$ (even)

(b) Check 1 $1+0+1+1$ (odd)
Check 2 $0+0+1+1$ (even)
Check 3 $1+1+1+1$ (even)

The error is in position $001_2 = 1$. So the first digit is wrong and the correct codename is 0001111. The message is 0111.

Exercise 3

1. $2^2 - 1 - 2 = 1$ and so there can only be 1 message bit.

Message group	Protected signal
0	000
1	111

In this case, the Hamming code is the same as R3.

2. Check 1 is on positions 1, 3, 5, 7, 9, 11, 13, 15; check 2 is on positions, 2, 3, 6, 7, 10, 11, 14, 15; etc.

Check 1	$0+1+0+0+0+1+1+0$	(odd)
Check 2	$1+1+0+0+1+1+1+0$	(odd)
Check 3	$1+0+0+0+1+1+1+0$	(even)
Check 4	$1+0+1+1+1+1+1+0$	(even)

The error is in position $0011_2 = 3$. The correct codename is 010100010111110 and the message is 00000111110.

3. (a) Channel encoding efficiency $= \dfrac{\text{Number of message bits}}{\text{Length of codename}} = \dfrac{2^c - 1 - c}{2^c - 1}$

Redundancy $= 1 - \text{efficiency} = \dfrac{c}{2^c - 1}$.

(b) As c increases, 2^c increases much more rapidly than c and so the efficiency becomes very high. With 10 check digits the efficiency is $\frac{1013}{1023} \approx 99\%$.

4.6 Continuous signals

For many purposes a top frequency of 8000 Hz is assumed. How often should the waveform be sampled?

Using the Sampling Theorem, the sampling interval should be not greater than $\frac{1}{16000}$ second or 62.5 μs.

Exercise 4

1. In decimal form the levels are 5, 2 (or 1), 8 (or 9), 11, 9, 12, 4, 5, 6, 2. This would be encoded as:

0101 0010 1000 1011 1001 1100 0100 0101 0110 0010.

2. The levels in decimal form are 11, 12, 7, 3, 8, 13, 7, 10, 1, 6.

5 The Mariner-9 code

5.1 Hamming distance

From the diagram find the three points p such that

$d(101, p) = 2$

The points are 011, 000 and 110.

(a) Identify the pairs of skew lines given by this quadruple.

(b) Find all other quadruples of 3-bit codenames such that the Hamming distance between any pair is 2.

(a) The lines 000 – 110 and 101 – 011 are skew.

(b) { 001, 010, 100, 111 } is the only other quadruple.

5.2 (n, k, h) codes

(a) Explain why the distance between any pair of 2-in-5 codenames must be at least 2.

(b) Find the minimum distance between any two codenames in Rh.

(a) If two codenames differed in only one position and one of them had exactly two 1's then the other must have either one 1 or three 1's, contrary to its definition. (A similar argument shows that a distance of 3 is also impossible.)

(b) The two codenames would differ in an entire block of h bits, one being 00 … 0 and the other 11 … 1. The minimum distance is therefore h.

What error protection is given by a code with minimum distance 5?

$h = 5$ is odd. According to the theorem, $\frac{1}{2}(5-1) = 2$ errors can be detected and corrected.

Classify the code consisting of all 4-bit codenames.

The digits of this code are all message digits and so $n = k = 4$. d (0000, 0001) = 1, for example, and so $h = 1$. It is a (4, 4, 1) code.

Use this method to find the other 7 codenames in the Hamming (6, 3, 3) code.

000000, 001110, 010101, 100011, 011011, 110110, 111000

Exercise 1

1. (a) Codenames have three digits, of which two are message digits. (Note that if you omit the final digit of each codename you have still four different codenames; the final digit is an even parity check digit). It may easily be checked that the distance between any pair of codenames is 2.

   (b) {001, 010, 100, 111}, already found geometrically . It could also be found by interchanging 0's and 1's in the even-weight code.

2. (a) (i) The (4, 2, 2) code : 0000, 0110, 1001, 1111

   (ii) The (8, 4, 4) code :

   00000000, 00011110, 00101101, 01001011,
   10000111, 00110011, 01010101, 10011001,
   01100110, 10101010, 11001100, 01111000,
   10110100, 11010010, 11100001, 11111111.

   (b) (i) The code can detect (but not correct) one error.

   (ii) The code can detect two errors but can only correct one error.

3. (a) (i) 10010 (ii) 00110.

   (b) (i) The error group would contain only one Y (in the position corresponding to the corrupted digit of the message group).

   (ii) The error group would contain only one N (in the position corresponding to the corrupted digit of the check group).

   (c) (i) YYYYN and NYNNN respectively.

(ii) In the first signal there is an error in the last digit of the check group. In the second there is an error in the second digit of the message group.

(iii) The machine could construct error words. If the error word contained just one N this could be disregarded, the message group being correct. If the error word contained just one Y the message group would be altered in the position indicated by this Y.

You may have gone on to consider the case of 2 errors. The situation appears similar except that the roles of N and Y are reversed. An error word containing just two Y's might indicate two errors in the check group; one containing two N's might indicate two errors in the message group. Unfortunately the case of an error in each group invalidates the use of this method of finding two errors and another way of finding the nearest code name has to be found.

5.3 Metric spaces

Exercise 2

1. (a) Calling the given codenames

$$\mathbf{u}_1 = 010, \ \mathbf{u}_2 = 110, \ \mathbf{u}_3 = 111,$$

the code consists of $\mathbf{0}$, $\mathbf{u}_1$, $\mathbf{u}_2$, $\mathbf{u}_3$ and

$$\mathbf{u}_1 + \mathbf{u}_2 = 100, \ \mathbf{u}_1 + \mathbf{u}_3 = 101, \ \mathbf{u}_2 + \mathbf{u}_3 = 001, \ \mathbf{u}_1 + \mathbf{u}_2 + \mathbf{u}_3 = 011.$$

This is a complete listing of C_3.

(b) There are several answers. Possibly the simplest are {100, 010, 001} and {001, 011, 111}.

(c) If the members are $\mathbf{v}_1$ and $\mathbf{v}_2$, then the complete set of linear combinations is $\mathbf{0}$, $\mathbf{v}_1$, $\mathbf{v}_2$ and $\mathbf{v}_1 + \mathbf{v}_2$; only 4 in all.

(d) If $\mathbf{r}$, $\mathbf{s}$ and $\mathbf{t}$ do not span C_3, then two of the eight linear combinations of $\mathbf{r}$, $\mathbf{s}$ and $\mathbf{t}$ must be equal. There are two possibilities:

two of $\mathbf{r}$, $\mathbf{s}$ and $\mathbf{t}$ are equal;

the sum of two of $\mathbf{r}$, $\mathbf{s}$ and $\mathbf{t}$ equals the third.

For example, 011, 110 and 101 do not span C_3 because

$$011 + 110 = 101.$$

2. (a) $0110 = 1100 + 0101 + 1111.$

   (b) This is not possible because:

$$0110 = a\,0100 + b\,0101 + c\,1011 + d\,1111$$
$$\Rightarrow 0 = c + d,\ 1 = a + b + d,\ 1 = c + d,\ 0 = b + c + d$$

$c + d$ cannot be equal to both 0 and 1.

This set of four codenames does not span C_4. You may have noticed that $0100 + 1011 = 1111$.

5.5 Spreading the message

(a) How can you distinguish between a randomly generated string of characters and one intended to convey a message?

(b) Messages sometimes appear to be periodic, for example SOS or 'Mayday' distress calls, which are repeated many times. Does this contradict the definition of messages given above?

(a) In a message you might expect the symbols to occur with varying frequencies. Failing this, digrams or other combinations should have different frequencies. If this is not the case then the stream must be randomly produced.

(b) You can think of the distress calls as simply the device to catch attention. In themselves, they contain no information specific to a particular incident.

TABLE 1
Logarithms to base 2

N	LOG_2N
1	0.000
2	1.000
3	1.585
4	2.000
5	2.322
6	2.585
7	2.807
8	3.000
9	3.170
10	3.322
11	3.459
12	3.585
13	3.700
14	3.807
15	3.907
16	4.000
17	4.087
18	4.170
19	4.248
20	4.322
21	4.392
22	4.459
23	4.524
24	4.585
25	4.644
26	4.700
27	4.755
28	4.807
29	4.858
30	4.907
31	4.954
32	5.000
33	5.044
34	5.087
35	5.129
36	5.170
37	5.209
38	5.248
39	5.285
40	5.322
41	5.358
42	5.392
43	5.426
44	5.459
45	5.492
46	5.524
47	5.555
48	5.585
49	5.615
50	5.644

N	LOG_2N
51	5.672
52	5.700
53	5.728
54	5.755
55	5.781
56	5.807
57	5.833
58	5.858
59	5.883
60	5.907
61	5.931
62	5.954
63	5.977
64	6.000
65	6.022
66	6.044
67	6.066
68	6.087
69	6.108
70	6.129
71	6.150
72	6.170
73	6.190
74	6.209
75	6.229
76	6.248
77	6.267
78	6.285
79	6.304
80	6.322
81	6.340
82	6.358
83	6.375
84	6.392
85	6.409
86	6.426
87	6.443
88	6.459
89	6.476
90	6.492
91	6.508
92	6.524
93	6.539
94	6.555
95	6.570
96	6.585
97	6.600
98	6.615
99	6.629
100	6.644

TABLE 2

Probabilities of symbols in English, in descending order, and Huffman codenames

Symbol	Probability	Codename	Symbol	Probability	Codename
space	0.1859	000	*F*	0.0208	001100
E	0.1031	101	*M*	0.0198	001101
T	0.0796	0010	*W*	0.0175	001110
A	0.0642	0100	*Y*	0.0164	011100
O	0.0632	0110	*G*	0.0152	011101
I	0.0575	1000	*P*	0.0152	011110
N	0.0574	1001	*B*	0.0127	011111
S	0.0514	1100	*V*	0.0083	0011110
R	0.0484	1101	*K*	0.0049	00111110
H	0.0467	1110	*X*	0.0013	001111111
L	0.0321	01010	*J*	0.0008	0011111101
D	0.0317	01011	*Q*	0.0008	00111111000
U	0.0228	11110	*Z*	0.0005	00111111001
C	0.0218	11111			

TABLE 3

Occurrence of English words

Word	Probability (x 10^{-4})
the	78.7
and	46.3
a	39.0
to	38.3
of	37.7
I	29.7
in	25.0
that	18.7
it	17.3
he	16.3
you	14.3
for	13.0
had	11.3
is	11.1
with	11.0
she	10.3
his	10.2
as	10.2
on	10.1
at	8.7
have	8.2
but	7.9
me	7.8
my	7.4
not	7.3

TABLE 4

Occurrence of pairs of letters

Pair	Probability (x 10^{-4})
E □	414
TH	379
□*T*	326
HE	294
□*A*	231
D □	225
T □	209
□*S*	169
AN	161
ND	153
HA	151
S □	139
N □	134
□*H*	127
IN	118
RE	118
□*W*	117
L □	115
□*O*	115
□*I*	110
LL	103

TABLE 5

International telegraph alphabet

Letter	Figure	Code Group	Letter	Figure	Code group
A		00111	*Q*	1	00010
B	?	01100	*R*	4	10101
C	:	10001	*S*	!	01011
D	Who are you?	01101	*T*	5	11110
E	3	01111	*U*	7	00011
F	%	01001	*V*	=	10000
G	@	10100	*W*	2	00110
H	£	11010	*X*	/	01000
I	8	10011	*Y*	6	01010
J	BELL	00101	*Z*	+	01110
K	(	00001	Carriage return		11101
L	)	10110	Line feed		10111
M	.	11000	Letter shift		00000
N	,	11001	Figure shift		00100
O	9	11100	Space		11011
P	0	10010			